ENGLISH GRAMMAR

LANGUAGE
AS HUMAN BEHAVIOR

Anita K. Barry

The University of Michigan–Flint

Prentice Hall, Upper Saddle River, New Jersey 07458

Library of Congress Cataloging-in-Publication Data

BARRY, ANITA K., (date)
    English grammar : language as human behavior / Anita K. Barry.
        p.   cm.
    Includes index.
    ISBN 0–13–835281–X
    1. English language—Grammar. 2. Human behavior. I. Title.
PE1112.B28 1998   97-10061
428.2—dc21   CIP

Editor in Chief: *Charlyce Jones Owen*
Acquisitions Editor: *Maggie Barbieri*
Editorial Assistant: *Joan Polk*
Production Liaison: *Fran Russello*
Editorial/Production Supervision: *Kim Gueterman*
Buyer: *Mary Ann Gloriande*
Art Director: *Jayne Conte*
Copyeditor: *Kathryn Beck*
Marketing Manager: *Rob Mejia*

This book was set in 10/12 Palatino by Compset, Inc.
and was printed and bound by Courier Companies, Inc.
The cover was printed by Phoenix Color Corp.

 © 1998 by Prentice-Hall, Inc.
Simon & Schuster / A Viacom Company
Upper Saddle River, New Jersey   07458

Printed in the United States of America
10 9 8 7 6 5 4 3 2 1

ISBN 0-13-835281-X

Prentice-Hall International (UK) Limited, *London*
Prentice-Hall of Australia Pty. Limited, *Sydney*
Prentice-Hall Canada Inc., *Toronto*
Prentice-Hall Hispanoamericana, S.A., *Mexico*
Prentice-Hall of India Private Limited, *New Delhi*
Prentice-Hall of Japan, Inc., *Tokyo*
Simon & Schuster Asia Pte. Ltd., *Singapore*
Editora Prentice-Hall do Brasil, Ltda., *Rio de Janeiro*

For my parents, Mildred and Jack Klitzner

and in memory of my brother, Barry M. Klitzner, M.D.

Remember when you go forth to seek your fame and fortune, it's not who you know that counts, it's whom.

Author Unknown

CONTENTS

PREFACE xi

1 WHY STUDY ENGLISH GRAMMAR? 1

Native Speakers and Grammar Study 1
Standard English 2
Judgments about English 4
The Legacy of the Eighteenth Century 7
Reflections 9
Notes 9

2 HOW DO WE STUDY ENGLISH GRAMMAR? 10

Why Do People Disagree about Grammar? 10
   Who Is the Authority? 10
   What Role Do Dictionaries Play? 10
   Why Is There No One Standard? 12
   Why Do Languages Change? 12
What Are the Common Elements of English? 14
   Constituent Structure 15
   Rules and Regularities 19
Reflections 20
Notes 21

3 NOUNS AND NOUN PHRASES 22

What Are Nouns? 22
What Are Some Common Subcategories of Nouns? 24
What Makes Up a Noun Phrase? 27
   Determiners 27
   Predeterminers and Postdeterminers 29

v

What Are the Functions of Noun Phrases?   30
   Subject   30
   Direct Object   32
   Indirect Object   33
   Object of a Preposition   34
   Complement   35
Verbal Nouns and Noun Phrases   36
Reflections   38
Practice Exercises   39

4   VERBS AND VERB PHRASES   43

What Are Verbs?   43
What about the Exceptions?   47
What Are Some Common Subcategories of Verbs?   50
What Is Verb Tense?   54
What Makes Up a Verb Phrase?   60
What Are Nonfinite Verb Phrases?   62
What Is Subject-Verb Agreement?   64
Reflections   69
Practice Exercises   71

5   PRONOUNS   77

What Are Pronouns?   77
Personal Pronouns   79
Reflexive Pronouns   85
Reciprocal Pronouns   88
Demonstrative Pronouns   88
Relative Pronouns   89
Interrogative Pronouns   91
Universal and Indefinite Pronouns   92
Reflections   93
Practice Exercises   96

6   ADJECTIVES AND ADVERBS   100

What Are Adjectives?   100
How Do Adjectives Modify Nouns?   102

What Are Adjective Phrases? 104
What Are Adverbs? 105
Is All Well and Good? 108
What Are Adverb Phrases? 111
Reflections 111
Practice Exercises 113

7 PREPOSITIONS AND PARTICLES 116

What Are Prepositions? 116
What Are Prepositional Phrases? 117
What Are Particles? 120
Reflections 122
Practice Exercises 123

8 LANGUAGE USERS AT WORK: MULTIPLE MEANINGS AND NEW CONSTITUENTS 127

Crossover Functions of Words: When Is a Noun Not
  a Noun? 128
Ambiguity: When Can a Sentence Mean More Than One
  Thing? 130
Creating New Constituents: How Do We Build New
  Structure? 133
Clause Coordination and Ellipsis: Adding and Subtracting
  for Efficiency 137
Reflections 138
Practice Exercises 141

9 CLAUSE TYPE: VOICE 145

What Is Grammatical Voice? 145
How Is the Passive Voice Formed? 147
How Are Grammatical Relations Determined in the Passive
  Voice? 148
Why Do We Need the Passive Voice? 150
What Is a Truncated Passive? 151
The Passive and Structural Ambiguity 153

Reflections   154
Practice Exercises   155

10   CLAUSE TYPE: DISCOURSE FUNCTION   159

What Is Discourse Function?   159
   Declaratives   160
   Interrogatives   161
   Imperatives   172
   Exclamatives   172
Crossover Functions of Clause Types   174
Reflections   176
Practice Exercises   177

11   CLAUSE TYPE: AFFIRMATIVE VERSUS NEGATIVE   182

What Is Negativity in Grammar?   182
Verb Negation   182
Negation of Indefinites   184
Noun Negation   186
Adjective and Adverb Negation   187
Negation of Compounds   188
Partial Negation   189
Reflections   191
Practice Exercises   193

12   COMBINING CLAUSES INTO SENTENCES   197

How Is a Sentence Different from a Clause?   197
Sentence Building through Coordination   197
Sentence Building through Subordination   199
Adverbial Clauses   201
Noun Clauses   202
Relative Clauses   207
Restrictive and Nonrestrictive Relative Clauses   209
Reduced Relative Clauses   211
Reflections   214
Practice Exercises   215

13 SENTENCE ANALYSIS 220

Simple Sentences 220
Compound Sentences 222
Complex Sentences 224
Compound-Complex Sentences 228
Reflections 230
Practice Exercises 231

14 WORD CONSTRUCTION, PRONUNCIATION, AND SPELLING 235

How Are Words Constructed? 235
What Is Phonology? 238
What Is Phonetics? 240
Some Rules of English Morphology 244
Variations in Pronunciation 246
Standard English and Spelling 249
Should We Reform Our Spelling System? 250
Reflections 253
Practice Exercises 255

ANSWERS TO PRACTICE EXERCISES 257

GLOSSARY 284

INDEX 295

PREFACE

This book is written for students of English grammar, who come to the task of studying the language with a variety of skills, interests, goals and expectations, not to mention fears and anxieties. It is addressed primarily to the native speaker of English, and so it is not designed to teach English. Rather, it builds on what students already know to develop an appreciation for how the language works.

The main focus of the book is on language as human behavior. Students are encouraged to view English not as an abstract system of rules, but as a product of people who seek patterns and regularity, who use language to communicate their needs and wishes and to exercise power over others, and who are capable of experiencing linguistic insecurity in the face of social judgments about their usage. Students also come to learn that the language they use is the product of people's use over a long period of time, not just a tool for the present, and that English, along with judgments about particular usage, shifts over time. The goal of the book is to make students feel that they are active participants in shaping their language rather than passive victims of grammar rules that someone imposes on them. They are encouraged to be curious about how others use English, and to be flexible enough to understand that there are competing descriptions of language structure, as well as competing opinions about correctness.

In its discussion of English, the book largely adheres to traditional grammatical terminology to keep continuity with our long and rich heritage of grammar studies. At the same time, the grammatical descriptions are informed by the insights gained from modern linguistic analysis. The merger of these two approaches gives students the necessary tools to think about how their language works without becoming entrenched in the mind-set of a particular theory. It also provides them with the flexibility to adapt to new terminology they might encounter elsewhere. Lastly, in keeping with the main goal of talking about human behavior, discussion turns to usage and usage questions wherever they are relevant.

The book is designed for a one-semester college course. It covers the basics of English without dwelling on the exceptional or the exotic. It begins with a discussion of the development of a standard English language and the origins of our present day rules of English and attitudes towards usage. Students are invited to explore their own recognition of standard

English and to appreciate that people may differ in their judgments. They learn that what is considered "correct" does not always match what sounds appropriate to them. The first chapter lays the foundation for the study of grammar, emphasizing the complex interaction between language rules and behavior. The second chapter talks about how one approaches the study of the structure of a language, including a brief discussion of how languages change over time. It also gives an overview of language structure, explaining the essentially hierarchical as opposed to linear nature of language. From there the book works from the lowest levels of grammatical organization to the highest, starting with an analysis of words and working up to the level of the sentence.

As students and teachers begin to work with this book, they will realize that the material is integrated in ways not apparent from the chapter headings. There is no part of language that is wholly separate from the other parts: it is an organic system in which the parts are interrelated and function together to perform the highly complex task of communicating human thought. Naturally, then, a description of a language cannot consist of wholly separate parts either. In this particular description, there is a good deal of recycling of information. Topics are not necessarily explored in their entirety when they are first introduced and may resurface in other contexts to have new light shed on them. In some cases a theme is introduced early and developed gradually throughout the book. The most commonly recurring themes have to do with the factors that influence people's use of their language: their common needs and preferences, and their shared strategies for turning their thoughts into words. Chapter 8 is the most evident example of the recycling design of the book. In this chapter, students are invited to take stock of all they have learned up to this point and to see how this information gives them a window into human linguistic creativity.

A special word needs to be said about the final chapter of the book. Grammar books often focus on syntax, the organization of words in sentences. Indeed, that is the primary focus of this book. But again, no part of a language exists in total isolation from the others, and so it makes sense to talk about the other things that make up the grammar of English—that is, the full set of rules and principles that govern its use. Chapter 14 is by no means intended as an exhaustive coverage of the phonetics, phonology, morphology and spelling system of English. But it does offer an opportunity for students to understand how all the parts of the language are interrelated and to recognize that certain recurring themes in syntax recur in other aspects of English as well.

Each chapter contains three types of exercises. First, there are short **Discussion Exercises** distributed throughout the chapter and designed for group work. These typically exemplify and reinforce a newly introduced principle. They give students a chance to check their own understanding in a nonthreatening forum and to spend part of every class period talking about language. On occasion these exercises are used to encourage students to extend what they have learned and to uncover new facts and principles of grammar themselves. In this way, the exercises become an integral part of the text material and serve a teaching as well as a review function.

Second, there are open-ended questions and project suggestions at the end of each chapter, called **Reflections**. These are intended to get students to think about language use—their own and others'—in real-life settings or to ponder some aspect of English structure that eludes analysis. These exercises are intended to stimulate further class discussion and to engage students in timely, enjoyable discourse about their language.

Finally, there are **Practice Exercises** at the end of each chapter that integrate all the information presented in that chapter. These are designed for students to work on outside of class. They are intended to be more closed-ended than the other exercise types, and they focus on purely structural material rather than on questions of usage. Answers to these exercises are provided at the end of the book. Of course, given the nature of language, these too often lend themselves to discussion.

The exercises taken together are designed to get students to think, talk, and write about English with increasing confidence and sophistication as the term progresses.

As anyone who has ever tried it knows, describing a language is an open-ended enterprise. There is always more that could be said. The goal here is to lay the necessary groundwork for thinking about language so that students can extend what they learn to new situations when the occasion arises, and to apply their knowledge in ways most useful to them, either in teaching the language to others, or in their own speaking and writing, or in making sense of the often subtle but always pervasive set of social judgments that accompany language use. This book is a conversation about English that approaches grammar as a process, not a product; and it is a book in which thoughtful explanations are valued over "correct" answers. Above all, it strives to stimulate excitement, enthusiasm, and wonder about English usage that will endure once the course is over.

ACKNOWLEDGMENTS

I have referred to this book as "a conversation about English." More accurately, it is a contribution to a conversation that has been taking place for hundreds of years among grammarians, linguists, English teachers, dictionary makers, and self-appointed guardians of the English language. This larger conversation is not an orderly one. There are differences of opinion and differences of approach, some minor and some major. Nevertheless, this collective thinking about the English language provides a rich and lively context in which to do one's own exploration. I gratefully acknowledge the work of the many other language scholars whose work has helped to shape my thoughts about English grammar and usage.

As with any particular work, there are some individuals whose contributions stand out above the rest. I wish to thank the editors at Prentice Hall, especially Maggie Barbieri and Kim Gueterman, for their ongoing help and support throughout the project. I am enormously grateful to the reviewers who forced me to clarify my thinking in more than a few places and who added a wealth of information and insight of their own: Nancy Hoar, Western New England College; Kitty Chen Dean, Nassau Community College; and Robin C. Barr, American University. Many of the examples and observations in the book are theirs. A special debt of gratitude is owed to my son, Michael Hochster, for his thoughtful reading of the manuscript and extremely helpful commentary, especially on questions and negation. I accept responsibility for any remaining errors or lapses of judgment. Also, I thank Bill Meyer, my husband, friend, and colleague, for his patient and engaged listening while I talked this book into being.

Finally, no acknowledgments would be complete without recognizing one other group of participants in this conversation about English : the students of Linguistics 244 at the University of Michigan–Flint. For over two decades I have been inspired by their wisdom and good humor. It is with great pleasure that I write this book for them.

Anita K. Barry
The University of Michigan–Flint

1

WHY STUDY ENGLISH GRAMMAR?

NATIVE SPEAKERS AND GRAMMAR STUDY

If you are a native speaker of English, the study of English grammar is not the same for you as the study of Spanish or Japanese grammar, for example. You already know English, and you know it well. You can construct complex sentences, ask questions, give orders, request help or permission, transmit information, express your feelings, make suggestions, scold, promise, apologize, accuse, warn, or say anything else that you want to say without a moment's hesitation. At this point in your life, you can also read and write English, with an even broader range of structures and vocabulary with which to express yourself. So what is to be gained from studying English grammar? What is there to learn that you don't already know? What you will learn in this book is that "knowing" a language can occur at several different levels. "Knowing" a language so that you can use it is not the same as "knowing" a language so that you can explain how it works. It is the purpose of this book to teach you how English works. Once you know that, you can explain it to others, or you can use it as a way of understanding and evaluating your own writing or speech.

Another reason people sometimes give for studying the grammar of English is to find out what is "correct." Speakers of English are sensitive to the social judgments that accompany variations in structure, vocabulary, and pronunciation. We are aware that not everyone speaks the same way, and, more importantly, we are acutely aware that some forms of English have higher social value than others and bring respect to their users. Linguistically speaking, no one form of English is better than another, but that does not change the social fact that some versions generally signal to us lower class and a lower level of education, while others signal higher class and a higher level of education. Thus it becomes important to all of us to understand how those judgments work and the social consequences of choosing one form of English over another. "Correct" versus "incorrect" English is much too simple a designation for what is in fact a complicated array of behaviors and social judgments. The idea of "correctness" is one of the important themes of this book. It is an idea that defies definition and only becomes clear to us as we explore how English is used and received in a variety of contexts. So, if learning "good grammar" is your primary motivation for studying English grammar, you will need to have patience, as the answers will unfold gradually as our conversation about English progresses.

STANDARD ENGLISH

No one can study all facets of English at once, although they are all worthy of study: the richness of its variation; the complexity of its pronunciation, vocabulary and sentence structure; its fascinating history; its spread around the world. All of these offer amazing insights into how people have used English for centuries to communicate with one another, to judge others, to exercise power over others, and to express their innermost needs and feelings. In this book we focus our attention on what is known as **Standard American English**. It is that form of English that is expected in public discourse in the United States: in newspapers and magazines, radio and television news broadcasts, textbooks, and public lectures. It is the form of English that is recognized as the English of the educated, irrespective of region, gender, or ethnicity. The written form of English, sometimes called literary Standard American English, does not vary as greatly as its spoken forms, even from country to country, so we can approach a universal description of the language by narrowing our focus primarily on its written form. As our discussion progresses, you will learn what constitutes Standard American English and how we determine whether a particular facet of English does or does not fall within this designation. Of course, we cannot explore these questions without also understanding something about the broader range of variation in English and the historical and social contexts in which it has developed and grown. English, like any language, is not merely a set of rules. It is the product of human behavior over many hundreds of years and will continue to evolve indefinitely into the future. The only reason we assign particular importance to twentieth-century Standard American English is that it is the form of English that has the most immediate consequences for us.

When we think about standard English, naturally we also think about *nonstandard English*. After all, if there were no variation in English, we wouldn't need to talk in terms of a "standard" at all. But all speakers of English recognize some forms of English as falling outside what we consider "correct" or "acceptable" or "normal" forms (although we may not always agree on just what those are). Sometimes we react to regional differences, which are largely differences in pronunciation and vocabulary. Some may think it curious that many speakers of English, including some New Yorkers and Bostonians leave out *r*s in some words when they speak, or that some southerners pronounce *pin* and *pen* alike. It may strike some of us as odd that people from other areas of the country pronounce *cot* and *caught* alike or that they stress the first syllable of *insurance*. You might be surprised to discover that what you call a *frying pan* other people call a *spider*, or what you call a *faucet* is a *spigot* to others. We experience a range of reactions to these differences, but overall they tend to be relatively mild and nonjudgmental. Furthermore, they are not likely to show up in the written form of the language—New Yorkers may say what sounds like *bawn* or *bahn* but they write *barn*. We make harsher judgments about language differences that we associate with social class. Many of these (but not all) involve sentence structure and may show up in the written as well as the spoken form of the language. A person who says *I ain't got none, I already seen it,* or *he done it hisself* may write those sentences that way as well. Whether spoken or written, utterances such as these trigger negative reactions from many people. Someone hearing such statements is likely to judge them as incorrect and associate them with lack of education and low social status, even though there is no linguistic foundation for these associations. As we will see in our later discussions, it is largely historical accident that leads to one particular form of English being favored over another. Nevertheless, however misguided these judgments may be, we cannot deny that they are widespread and can affect our lives in many ways. That is why we often experience some degree of linguistic insecurity and worry about whether some things we say are standard or nonstandard, correct or incorrect.

Interestingly, our reactions to nonstandard English are not always negative. In some cases, people hypercorrect—that is, in an attempt to avoid what they know to be a grammatical error, they overapply or misapply a rule. *Between you and I* is a common example of such **hypercorrection**. Hypercorrections tend to occur occasionally in otherwise standard usage, so they often go unnoticed or do not characterize a person's usage as uneducated. Furthermore, they tend to be used by people who, for reasons other than language, carry higher social status or wish to. Hypercorrections, although nonstandard, may even signal higher status, at least to those who are themselves unsure of the standard forms.

DISCUSSION EXERCISE 1.1

Each of the sentences below violates some rule of strictly formal standard English. By the time you reach the end of this book, you will understand what makes them nonstandard. For now, try to anticipate what your reactions might be to someone who said them. Would you have the same reaction in each case? Which make you think that the speaker lacks education? Do any sound fine to your ear? Do any make you think the speaker is well educated? Why do you think your reactions might change from sentence to sentence, even though they are all nonstandard? Do different people register different reactions to the same sentences?

Kathy and me arrived first.

I ain't seen 'em.

Whom shall I say is calling?

It don't matter to me.

We been here a long time.

That man was kind to my sister and I.

Someone left their umbrella here.

We was right.

If the weather was warmer, we could have a picnic.

They did it theirselves.

I'm right, aren't I?

That's me in the photograph.

Nobody gets nothing for free.

She is smarter than him.

JUDGMENTS ABOUT ENGLISH

You might be surprised to learn that speakers of English did not always make such strong judgments about others' use of the language. Before the printing press was introduced into England, where English began, there was no recognized standard, nor was there any real need for one. Much of the important public writing was either in French or Latin. English was used primarily for oral and informal purposes and varied quite a bit from place to place, with especially large differences between the north of England and the south. William Caxton, who introduced the printing press into England, noted in his writing in 1490:[1]

> And certainly our langage now used varyeth ferre from that whiche was used and spoken whan I was borne. For we englysshe men ben borne under the domynacyon of the mone, whiche is never stedfaste, but ever waverynge,

wexynge one season, and waneth & decreaseth another season. And that comyn englysshe that is spoken in one shyre varyeth from a nother.

When people did write English, they had no common spelling system, so you would see the same word spelled different ways from one author to the next, and sometimes even in the work of one author. For example, in one fourteenth-century poem we see "English" spelled both "English" and "Englysch," both different from the spelling in the passage cited above from Caxton's work.[2]

After 1476, the year the printing press was introduced into England, lack of a written standard became a practical concern. Now England had the technological capacity to spread the written language to large numbers of people throughout the country, but one couldn't count on a large reading audience unless there was a shared written language. What would that be? The natural choice for this standard was the form of English used in London. In some ways, it represented a compromise between the north and the south, sharing some of the features of each. It also enjoyed prestige throughout England. London was a prosperous city and was generally recognized as a center of learning, with both Oxford and Cambridge Universities nearby. And so, by general consensus and without decree, London English became the model to which others looked for use as a formal standard. Of course, there were authors who used other varieties of English as well, and there was still a great deal of variation as writers and printers sorted out how they would represent English on paper.

What is not in evidence during this period of development of English is a sense of condemnation of those who did not use the standard. But with the coming of the eighteenth century, the Age of Reason, we see a dramatic shift in attitude among a small but influential group of writers and scholars. Believing that language ought to be unvarying and permanent, like Classical Latin and Classical Greek (which were no longer spoken), and logical, they were appalled at what appeared to them as chaos in English. They thought it was wrong that people who spoke English invented new words and phrases, shortened others, borrowed words from other languages, allowed the meanings of words to change, and expressed the same grammatical idea in more than one way. For example, one source of concern to them was the fact that some people said *My brother is taller than I* while others said *My brother is taller than me.* They thought English should be "pure": without variation from one person to the next, without change over time, without irregularities, and without contamination from other languages. They apparently believed that English *could* be pure, if not for the corrupting influence of the people who used (and "abused") it. In their fervor, they sought to have an academy established in England, a governmental body that would officially regulate use of the language. Such academies were already in operation in other European countries, such as France and Italy. But the proposal, among whose greatest supporters was Jonathan Swift, the author of *Gulliver's Travels*, failed to gain the necessary support in the English parliament and eventually died.

The failure to establish an academy was a blow to its supporters, but rather than accept defeat, they responded with linguistic entrepreneurship. Individuals set out to write dictionaries and grammar books designed to achieve the same purposes. These publications, by setting down rules for the English language, would tell the English (and later the Americans) what was right and what was wrong once and for all. Among those early grammarians and dictionary makers (lexicographers) were Samuel Johnson, Robert Lowth, and Noah Webster. Noah Webster, of course, was most influential in establishing an American standard, somewhat different in detail from the British standard, and we still see his name associated with some of our current dictionaries.

It is interesting to consider how the eighteenth-century grammarians decided what the rules of English ought to be. Actual usage could not be their guide, since for them that was the source of the problem. They certainly were not to be swayed by the preferences of ordinary people using the language in ordinary ways. Believing that language systems were by nature logical, they relied on logic to help them make decisions among competing usages. Remember the example mentioned earlier: should we say *My brother is taller than I* or *My brother is taller than me*? In this case they reasoned that there was an implied continuation of a sentence that we do not actually say: *My brother is taller than __ am tall.* Continuing the sentence in our heads would tell us that the correct choice is *I* rather than *me*.

In other cases they looked to Classical Latin and Greek as models for correct structure. These languages existed only in written form as of the eighteenth century. They were not used any longer by people for everyday conversation, and so they had the stability and uniformity that the English grammarians craved for English. What did they say about the choice between *This is her* and *This is she*? You might be able to guess if you knew Latin, which has a rule that says "after a copular verb use the nominative case." If you apply the rule to English, *This is she* becomes the correct choice.

The grammarians also based their judgments on English history. Since they viewed language change as the equivalent of language decay, they tended to assume that earlier forms and meanings were correct, while the more recent ones were wrong. For example, using the criterion of history, they declared that *demean* should mean "behave" (as it still does in *demeanor*), rather than its later meaning of "debase."

From this brief description, you can see that the foundations of formal English grammar were based on reasoning (often flawed) and introspection, not on actual usage. What was current and popular was not necessarily considered correct by most grammarians. By the nineteenth century, views of grammar had begun to shift and more value was placed on usage and the collection of information about actual usage. *The Oxford English Dictionary*, for example, begun in 1879 and finally published in 1928, recorded the history of words based on their occurrence in writing over many centuries and gleaned the meanings of words from the contexts in which they were used.

Then in the twentieth century the developing field of linguistics argued convincingly that all living languages change over time, that variation is natural and inevitable, and that all grammatical systems are equally capable of expressing whatever it is that people want or need to communicate. It is also a fundamental principle of linguistics that the structure of each language is to be described on its own terms and not forced into the mold of another language, such as Latin. The related field of sociolinguistics in particular recognizes the scholarly benefits of analyzing the grammars of different varieties of a language rather than trying to suppress them as inferior forms of the one designated as standard.

DISCUSSION EXERCISE 1.2

1. One of the better known grammar rules that emerged from the eighteenth century deliberations on English is the rule about negatives: you may not have two negatives in the same sentence. Thus, although double negatives such as *I don't have no money* were common in earlier forms of English and occur in many other languages as well, they were banned from standard English in the next century. Why do you think the grammarians frowned upon them?

2. You will remember that the grammarians decided that *He is taller than I* is correct, while *He is taller than me* is not. Can you think of an equally plausible argument for choosing the second sentence as the correct one?

3. Standard English grammar requires that we say *different from* and not *different than*. How do you think grammarians arrived at this conclusion? (Hint: think of the verb *differ*.)

4. What do you think eighteenth-century grammarians might have said about Shakespeare's *most unkindest cut of all*?

THE LEGACY OF THE EIGHTEENTH CENTURY

Although in more recent centuries language scholars have recognized the realities of language usage, we are still left with some of the attitudes about English that dominated the thinking of the eighteenth-century grammarians. We still assign particular value to standard English and see it as better than other varieties of English. We devote a considerable portion of our educational resources to teaching people how to read and write it. We regard knowledge of it as a prerequisite to many professions and question the abilities of people who have not mastered it. As a society, we are not especially accepting of variation in usage and tend to assume that one way is the "right" way while the others are "wrong." How many times have you asked yourself questions like: which is right, *This is the man who I met* or *This is the man whom I met*? It is rare for someone to assume that they are both right. We voice similar questions about pronunciation. Which is it, *harássment* or *hárassment*? To reinforce this notion of "one right answer," we have modern-day grammatical purists, such as John Simon and Edwin Newman, who write best-selling books telling us grammatical right from wrong.[3]

We also tend to assume that somewhere "out there" lie the answers to all our questions, just as the eighteenth-century grammarians assumed that all grammatical differences could be explained away by some principle of logic or an appeal to an older language. It is just a matter of finding the answer that we need. Sometimes we will consult someone whom we view as an authority on the language (an English teacher, for example) or enroll in a course in English grammar. In other cases we may consult a grammar book in the library or look something up in the dictionary. As we will see in the next chapter and all the chapters after that, getting satisfying answers to our grammar questions is never that simple because English is not Classical Latin or Greek; rather, it is a dynamic, living language, the product of the many millions of people who use it.

Before we turn to questions about how we study grammar, we might ask one more question about English usage. If, as a society, we generally recognize that standard English is highly valued, why doesn't everyone speak it and write it? In other words, why does variation in English persist? As you might guess, this too is a complex question with no easy answer. But we can suggest some ways of approaching an answer. In some cases, people are simply not presented with some ready model of standard English and might not be fully aware of the extent to which it is valued, or if they do know, they might not have enough exposure to it to model their own language after it. In other cases, people might recognize that other people use standard English but might not see it as appropriate to their own circumstances. For many, standard English carries with it an aura of formality, even stiffness, that makes it inappropriate in intimate or casual settings or in some work settings. In those instances, the less formal non-standard variety is more valued and signals a person as part of the group, an insider. Think about the lyrics of popular songs, for example. How often do you hear *ain't* or double negatives? Often people need to choose between grammatical correctness and appropriateness. If you see your best friend lingering after class, would you ask *For whom are you waiting*? or *Who are you waiting for*? (Ignore the question if your best friend is a grammatical purist.) Like forms of dress, different forms of English are appropriate for different circumstances.

In the next chapter we will talk about how we approach the study of English. Our goal is to focus our sights so that we come away with a coherent picture of how the language works despite the complexity that naturally accompanies any discussion of human behavior.

REFLECTIONS

1. One eighteenth-century grammarian defines grammar as "the art of speaking and writing any language with propriety. An art is a rational method, a system of rules, digested into convenient order, for the teaching and learning of something." Is this how you would define *grammar*? Is this how you would define *an art*?

2. Show the sentences of Discussion Exercise 1.1 to several different people whose command of English grammar you respect. Ask them to tell you which are not standard English. Did you get some disagreement among them? Did anyone say they are *all* not standard English? Why do you think there are differences of opinion about these sentences (assuming that there are) even among speakers of standard English? (A word of caution: educated people can be *very* touchy about grammar!)

3. Can you think of a situation in which you had a grammar disagreement with someone? How did you resolve it?

4. The French academy periodically tries to purge French of its English borrowings, such as *le weekend*. How successful do you think it has been in keeping French "pure"?

NOTES

1. From Caxton's preface to his *Eneydos*, as quoted in A. C. Baugh and T. Cable, *A History of the English Language*, 4th ed. (Englewood Cliffs, NJ: Prentice Hall, 1993, p. 191). Here is a translation:

   And certainly our language now used varies far from that which was used and spoken when I was born. For we Englishmen are born under the domination of the moon, which is never steadfast, but ever wavering, waxing one season, and wanes and decreases another season. And that common English that is spoken in one shire varies from another.

2. William of Nassyngton's *Speculum Vitae* or *Mirror of Life* (c. 1325) as quoted in Baugh and Cable, p. 141.

3. See, for example, Edwin Newman's *Strictly Speaking* or *A Civil Tongue*; or John Simon's *Paradigms Lost*.

2

HOW DO WE STUDY ENGLISH GRAMMAR?

WHY DO PEOPLE DISAGREE ABOUT GRAMMAR?

Who Is the Authority?

In the preceding chapter we discussed how many of our rules of English grammar were handed down to us from the eighteenth-century grammarians, who based their decisions about right and wrong largely on logic, history, or comparison to Classical Latin and Greek. For some people today, those rules are the final word about correct English. But most of us do not rely heavily on books that were written two hundred years ago to tell us about English today. Rather, we take a more practical view of language use and look for cues in our contemporary lives to guide us in the use of standard English. We look for models of what we regard as standard usage, and we consult contemporary sources, including teachers, editors, dictionaries, and grammar handbooks. But we still find that getting answers is not so easy as it might seem at first. If we had an academy, perhaps the problem would be less troublesome. At least there would be a unique authority that everyone could consult, and differences of usage and opinions about usage might be resolved in a fixed and predictable way. But we do not have an academy, nor do we have any other special authority recognized by everyone as the last word on English usage. Instead, we have lots of different sources, and by *sources* we mean real people who are faced with decisions, just as the eighteenth-century grammarians were.

What Role Do Dictionaries Play?

Let's take a closer look, for example, at the task of publishing a dictionary of English. Suppose you decided to publish one. How would you decide what meanings and pronunciations of words to include? Would you rely on older uses? Would you rely on the judgments of a few well-educated and influential scholars? Would you try to sample a wide range of people in different walks of life and list the most common usage? Would you rely only on written documents as sources of information? There are no right answers to these questions and, in fact, different dictionary makers have different answers to them, so that dictionaries themselves may differ in their purposes and their methods of making decisions. *Webster's Third New International Dictionary*, for example, attempts to reflect actual usage in neutral, descriptive terms, omitting designations such as *illiterate*. These values are articulated in the preface to the dictionary. In the words of editor-in-chief Philip Gove:

> Accuracy in addition to requiring freedom from error and conformity to truth requires a dictionary to state meanings in which words are in fact used, not to give editorial opinion on what their meanings should be.[1]

About pronunciation, he says:

> This edition shows as far as possible the pronunciations prevailing in general cultivated conversational usage, both informal and formal, throughout the English-speaking world. It does not attempt to dictate what that usage should be.[2]

Another widely used dictionary, *The American Heritage Dictionary of the English Language,* Third Edition, leans more toward representing educated speech only and relies on the judgments of a usage panel made up of writers, editors, and scholars, including professors of English and linguistics, and others who "occupy distinguished positions in law, diplomacy, government, business, science and technology, medicine, and the arts."[3]

Suppose you wanted to check the status of the word *ain't*. *Webster's Third International* tells us

> . . .though disapproved by many and more common in less educated speech, used orally in most parts of the U.S. by many cultivated speakers esp. in the phrase *ain't I*. (p. 45)

The American Heritage Dictionary, on the other hand, says:

> The use of ain't . . . has a long history, but ain't has come to be regarded as a mark of illiteracy and has by now acquired such a stigma that it is beyond any possibility of rehabilitation. (p. 37, Third Edition)

If you want to use a dictionary as a guide to your own usage of *ain't*, or as a means of judging the usage of others, then you will have to decide which of these accounts to rely on. And, of course, there are other dictionaries on the market as well, and each has its own approach to representing English. As discouraging as this might be to those of us who want definitive answers, the reality is that there is no unique authority on our language, and looking up a word in "the dictionary" is a comfortable fiction. In reality, we are looking up a word in "*a* dictionary."

DISCUSSION EXERCISE 2.1

1. Before 1961, Webster's dictionaries were more prescriptive in their approach to English usage—that is, more inclined to dictate correct usage. When the *Third International* announced its new policies in 1961, many people reacted with outrage. What do you think prompted this reaction? What do you think your own reaction might have been?
2. Which is standard English, *He has swam a mile* or *He has swum a mile*? What do you think the difference is in the way *Webster's* and *American Heritage* convey information about their use?

Why Is There No One Standard?

Another reason we have difficulty fixing on just one "correct" English is that modern English is spoken all over the world by hundreds of millions of people, and so perceptions of correctness will vary, even among the most educated and influential. As English spreads, it develops different standards. As we noted in the first chapter, Noah Webster succeeded in distinguishing an American standard from a British standard, an important step in the development of an American national identity. So now we recognize that there may be two acceptable ways to spell a word—*check* or *cheque, center* or *centre*, for example. Similarly, there are two standard pronunciations for some words, such as *schedule* and *lieutenant*. Or, it might be equally correct to say *the team is playing* (American English) or *the team are playing* (British English). British and American English are the two most influential standards around the world, but we must remember that each English-speaking country develops its own. Therefore, we should expect to find standard forms of English specific to Canada, Australia, and New Zealand as well. It is also the case that English is now used as one of the major languages in many countries of Africa and Asia and is developing standards specific to those areas. What was originally a language spoken by a few million people on one small island in Europe has now become a world language with many different varieties, and with identities separate from either British or American English.

Why Do Languages Change?

To complicate the picture still further, we have to keep in mind that languages change over time, and along with changes in language come changes in judgments about language. That is, "correct English" is a moving target. What was considered correct a hundred years ago is not necessarily what is considered correct today. The eighteenth century grammarians argued that English could be perfect and permanent if not for the laziness and carelessness of its users. For them, change was the equivalent of language decay. But modern linguists argue that change is inherent to all languages; without the flexibility to change, languages would not be able to serve the continuously evolving needs of the people who use them. If English had not been able to change, you would not have the words to talk about your hard drive or your floppy disk or even your carburetor! Language users are receptive to the enrichment of added vocabulary, while they shed words that are no longer of use to them. When was the last time you heard someone talk about their *trousers* or *breeches* or their *icebox* and *phonograph*? Do you sit on a *davenport* or keep your clothes in a *bureau*?

In addition to shifts in vocabulary, there is an even more important facet of language change to which we are all particularly sensitive: changes in our grammatical system. Grammatical systems are based on rules, or patterns. As people learn their language as children, they learn these patterns. For example, children learning English figure out that to make a noun plural, you have to add the suffix -*s*, or to express a past action, you must add the suffix -*ed* to a verb. But it is also true that there are exceptions to these patterns, sometimes because words remain unchanged from earlier times, when other patterns held, or sometimes because we borrow words into English from languages with different patterns. So, for example, *boy* fits the regular pattern for noun plurals (*boys*), while *man* and *crisis* do not (*men*, not *mans*; *crises*, not *crisises*). *Talk* fits the regular pattern for the past tense (*talked*), but *buy* does not (*bought*, not *buyed*). Unlike words that fit the regular patterns, exceptions are hard to learn. We have to learn them one by one and we have to remember each one separately. We need to hear them frequently for the irregularity to become fixed in our memories. When we look at how English has evolved since its beginnings, we see that collectively in our use of the language we strive to eliminate the irregularities by changing them to fit the normal pattern. If you look at earlier forms of English, you will find that *shoes*, for example, used to be *shoon*, and *eyes* used to be *eyen*; *climbed* used to be *clomb*, and *helped* used to be *holp*. Although no one person decides to make a change, together over the years we have changed English a great deal, so that many more nouns and verbs now fit the regular pattern. What this tells us is that language users can detect patterns easily and, from a broad historical perspective, prefer to have words fall within the patterns rather than outside them.

Clearly then, some words that are considered standard at some point in the history of English will drop out and be replaced by their regularized counterparts. Most of us can accept that without difficulty; we don't expect even the most educated among us to sound like Chaucer or Shakespeare. But what some of us find hard to accept is that English continues to change.

It is a dynamic, living system forever being shaped by the people who use it. The preference for regularity is no less compelling now than it was two hundred or more years ago, and people's linguistic behavior is no different from the way it has always been. Nevertheless, it is one thing to observe language change from a comfortable distance; it is quite another to experience it yourself. The first is often an interesting academic exercise, while the second can be disconcerting or even disturbing. Consider, for example, your reaction to someone who says *I knowed it*. Intellectually, we can register this as merely another example of regularization of the past tense. At the same time, for many of us it also signals lack of education. But as we know from observing the history of English, many regularized forms do take hold over time and come to be regarded as standard and educated.

How does that transition take place? How do we know when a newer form has replaced an older form? How do we know when it is no longer a stigma to use the newer form? Where's that academy when you need it? This is the source of grammar anxiety for many speakers of English. When a newer form is replacing an older form, they may both be used for a long time. It is only gradually that the older one will drop out. Meanwhile, we hear both being used. The ghosts of the eighteenth-century grammarians whisper to us that if there are two forms, one must be wrong. Our own experience tells us that regularized forms are stigmatized when they are first introduced. So we want to know when a word has achieved acceptability. (This could apply to grammatical constructions as well, as we will see later in the book.) But only our collective judgment determines that, so individually we often cannot get the immediate answers we seek. Should we say *dreamt* or *dreamed, lit* or *lighted, I have proven the theorem* or *I have proved the theorem*?

WHAT ARE THE COMMON ELEMENTS OF ENGLISH?

When we study the grammar of English, we have to take all of this into account: the absence of a unique authority, the variety of standards that exist around the world today, and the fact that English is continuously evolving and so are judgments about usage. This makes the study of English grammar an exciting challenge, but not impossible. As we said in Chapter 1, we need to focus our efforts so that we aren't trying to do everything at once. In this book, we focus on Standard American English. Even that, as we now know, is no simple exercise. We have to be flexible in our approach, attentive to the fact that we are talking about real people and not abstractions, and accepting of the idea that the answers to our questions may come in the form of thoughtful discussion rather than labels of "correct" and "incorrect."

Most importantly, however, we need to recognize that for all its variation and for all the indeterminacy in defining it, English is still English. People who speak it in all its varieties can understand one another and share the same written language, more or less, and share the same written language. English, like all languages, must meet the communication needs of the people who use it. That means no matter what variety of English we speak it at least must allow us to identify and make reference to things, to people, and to ideas. It must be able to describe actions and tell when they happened. It must allow us to give descriptions of things, people, and ideas. It must allow us to give information and to get information, to give orders, to express our feelings, to indicate relationships among things, people, and ideas, and to combine simpler ideas into more complex ideas. All forms of English meet these expectations and do so in similar ways. The varieties of English are more alike than different, and their common elements make English distinct from other languages. The rest of this book will concentrate on the common elements, using Standard American English as the focus of attention and basis for comparison to other varieties.

Constituent Structure

One feature common to all forms of English is that they have *constituent structure*. When we hear English, it seems to us that words just come out one after the other, like beads on a string. But, as we will see as we begin to examine the language, they are organized so that some elements bear a special relationship to each other that excludes others. For example, if you look at the sentence in (1) you will see that it consists of ten words.

(1)   The excited child chased the new puppy around the garden.

But you will notice that some of these words seem to group together and may stand alone in conversation as an answer to a question about this event. For example:

Who did it? *the excited child*
What did she chase? *the new puppy*
Where? *around the garden*
Around where? *the garden*
Did what? *chased the new puppy around the garden*

If you were asked to draw lines separating the parts of the sentence, you would probably insert them after *child, puppy,* and *garden.* Our mind simply tells us that certain words group together. Notice that there are other words that appear in sequence also, but they do not constitute a grouping. There is no question that could be answered *child chased the* or *new puppy around.*

Nor would we separate off those words together as groupings according to our intuitions. The groupings that hold together are called **constituents**. Constituents can be very short, like *rice* in sentence (2) or very long, like *because she knew that her life would be in danger if she revealed her sources to the FBI* in sentence (3).

(2)   *Rice* is high in carbohydrates.

(3)   The reporter refused to speak *because she knew that her life would be in danger if she revealed her sources to the FBI.*

Furthermore, you have already seen in sentence (1) above that constituents can nest inside other constituents. In other words, constituents are arranged hierarchically as well as linearly. So, for example, the constituent we have identified in (3) contains constituents within it: *her life, in danger, her sources, to the FBI.* And, *to the FBI* itself contains the constituent *the FBI.*

DISCUSSION EXERCISE 2.2

1. Identify some constituents in each of the following sentences. Judge what feels like a group to you and then see whether it could stand alone as an answer to a question in a conversation. Remember that constituents can nest inside larger ones.

    The bored students ignored the teacher's questions.

    They laughed hysterically when Kramer entered the room.

    The fact that the speaker showed up late annoyed many members of the club.

    Skiing in the Alps is my favorite vacation.

    The baby crawled into the closet and fell asleep.

2. We might show how one constituent is included within another by using brackets, as in the following: [to [the FBI]]. Place brackets around the constituents of *the man in the white coat.*

Constituents, or groupings, occur at many different levels of English, from the lowest level of the **root** and the **affix**, to the **word**, the **phrase**, the **clause**, and the **sentence**. In this book we will work our way from the lowest to the highest constituents. *Roots* and *affixes* (the more general term for **prefixes** and **suffixes**) are the components of words. For example, the word *cats* consists of the root *cat* and the suffix *-s*; the word *talked* consists of the root *talk* and the suffix *-ed*; *redo* consists of the root *do* and the prefix *re-*. Affixes are of two types: **inflectional** and **derivational**. *Inflectional affixes* express some grammatical information, like plural, or past tense, and do not change the basic category of the root. English has a small number of these, and when they occur, they help us to identify the category of the root. We know *talked* is a verb, for example, because it has the inflectional affix *-ed*, which gets attached only to verb roots. *Derivational affixes*, on the other hand, usually change one category into another. *Educate* is a verb; if I add the suffix *-tion*, it turns into the noun *education*. The suffix *-tion* is an example of a derivational affix in English and, as you might guess, there are many more derivational than inflectional affixes.

DISCUSSION EXERCISE 2.3

1. All the inflectional affixes in English are suffixes. The most common ones signal

   the plural *-s*

   the possessive *-s*

   the third person singular present tense *-s*

   the past tense *-ed*

   the past participle *-ed*

   the present participle *-ing*

   the comparative *-er*

   the superlative *-est*

   You might not be familiar with all this terminology yet, but try to pick out one example of each of these in the following sentence:

   > The man's son decided that he was leaving home because he had wasted all this time shoveling sidewalks and now he wanted to live in a warmer climate where the lowest temperature was 50°.

2. All affixes that aren't inflectional are derivational. English has many derivational prefixes and suffixes. For each one given, list several more words that use the same affix. Notice that the suffixes typically change the grammatical category of the word, while the prefixes do not.

*un*happy	govern*ment*
*dis*connect	seren*ity*
*re*read	equal*ize*

The next level of grammatical structure, as we have already implied, is the *word*, the result of putting roots and affixes together. Some words are just roots; others are combinations of roots and affixes. Words fall into different categories depending on their meanings, their functions, and the kinds of affixes they have. We refer to these categories as **lexical categories**, **word classes**, or **parts of speech**. They have names that are familiar to most people: *noun, verb, adjective, adverb, pronoun, preposition*, and *article* are some of the most common. Many of these word classes also have subcategories. You probably know the difference between a *common noun* like *boy* and a *proper noun* like *Bill*. You might also know the distinction between a *transitive verb* like *buy* and an *intransitive verb* like *laugh*. Do you know the difference between a *relative* and a *reflexive pronoun*? A *gradable* or *nongradable adjective*? If not, you soon will.

Words group together at the level of the **phrase**. A phrase has one part of speech at its core, called the **head** of the phrase, which gives the phrase its name, such as *noun phrase* or *verb phrase*. The phrase also includes all the other things that go with the head to form a group. These additional elements are called **modifiers**. If you look again at sentence (1) above, you will see that all the constituents we identified happened to be phrases.

DISCUSSION EXERCISE 2.4

Look again at sentence (1): The excited child chased the new puppy around the garden.

1. Find:

> three noun phrases
>
> one prepositional phrase
>
> one verb phrase

2. Tell what the head and modifiers are of each phrase you identify.

Phrases may occur together to make larger groupings, of course. The combination of a noun phrase followed by a verb phrase has special status: it is called a *clause*. The noun phrase and the verb phrase of the clause are also referred to as the **subject** and the **predicate** of the clause. Some clauses can stand all by themselves and are called **independent clauses**; others must attach to another clause and are called **dependent clauses**. Clauses may then combine into a larger constituent called a *sentence*.

All forms of English operate at all of these levels simultaneously, which sometimes makes it difficult to talk about one level without talking about the others. The figure below may help you to visualize the hierarchical structure of English that we have just described.

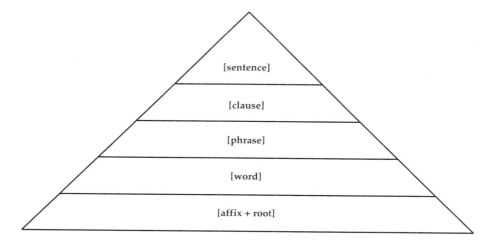

Rules and Regularities

Also common to all forms of English are rules that express patterns in the language. Rules may vary somewhat from one variety of English to the next, but most of them are the same, which is what gives the language its continuity. One kind of rule expresses the linear order in which elements must occur within their constituents. An example of such a rule is: "adjectives precede the nouns they modify." We all know that no one would say *I caught the ball red*, for example. We take that for granted, but we must keep in mind that this rule is one of the things that makes English different from, say, Spanish or French. Another kind of rule in English grammar expresses what elements can occupy the same constituent, that is, what elements are allowed to group together. Shared knowledge of the rules of acceptable grouping is what allows us to make the same judgments about what does and does not make up a constituent. Finally, there are rules for English that express relationships between elements, sometimes within one constituent, sometimes across constituents. We call these **cross-referencing rules**. One such cross-referencing rule for English is: "pronouns must agree in gender and number with their antecedents." You may not be familiar with the terminology, but if someone says *the girls hurt himself*, you know something is wrong!

DISCUSSION EXERCISE 2.5

What type of rule is each of the following: linear order, grouping, or cross-referencing? Can you translate them into ordinary English by explaining what we don't say?

Verbs agree with their subjects in person and number.

Single-word modifiers precede the head noun in a noun phrase.

Transitive verbs require an object noun phrase.

Prepositional phrases are made up of a preposition and a noun phrase.

English also has exceptions to its rules. There are parts of English that do not conform to regular patterns and do not lend themselves to generalization. We have already seen some examples of those: nouns that do not add the suffix *-s* to form the plural, verbs that do not add the suffix *-ed* to form the past tense. As we said earlier, sometimes these irregularities are holdovers from older patterns and sometimes they are borrowed from other languages. They are often the least stable part of the language because people prefer regularity in their grammatical systems. They are the most interesting part of the language as well, because individuals approach the problems they present in different ways, giving rise to variation in usage.

In the chapters to come we will embark on a careful examination of English, from the lowest to the highest levels of grammatical organization.

We will talk about the categories that make up each level and describe the rules for organizing them into acceptable patterns. We will also talk about the people behind the rules: How do we react to the irregularities in our grammatical system? What happens when standard English is inefficient or makes it difficult for us to express what we need to express? Observing people's language behavior gives us insight into how people organize a complex system of information in their minds and apply it in their everyday lives to communicate with others.

DISCUSSION EXERCISE 2.6

You have just been presented with much of the grammatical terminology that we will use to describe English in the rest of this book. At this point, it is certainly not expected that you will understand exactly what each means nor how it is applied in grammatical description. But we can use these terms for a "warm-up." As native speakers of English, we all have intuitions about the structure of the language. We know what sounds right or complete; we know what makes sense and what doesn't. What you need to begin to do now, as students of grammar, is to couch your descriptions in grammatical rather than subjective terms. Imagine, for example, that someone who doesn't know English very well has made the following statements. Describe what is non-English about these, using as much as possible the grammatical terminology we have introduced and avoiding impressionistic judgments like "it just sounds funny."

What your name is?

The Mr. Smith wrote this letter.

My friend taked the train to work.

These book is difficult.

They laughed the girl.

When I get to school.

Little boy caught ball.

The woman in the courtyard.

This is the most biggest package.

He a hat bought.

REFLECTIONS

1. If you would like to know more about dictionaries and dictionary making, the following sources will be of interest to you:

> Bryson, Bill, "Order Out of Chaos," *About Language: A Reader for Writers* (3rd ed.), pp. 184–96, ed. William H. Roberts and Gregoire Turgeon. Boston: Houghton Mifflin, 1992.

Gove, Philip B., "The Dictionary's Function," *Dartmouth Alumni Magazine* (May 1962), 10–11. Reprinted in *Focusing on Language: A Reader*, pp. 239–44, ed. Harold B. Allen, Enola Borgh, and Verna L. Newsome. New York: Thomas Y. Crowell, 1975.

Marquardt, Albert H., "The New Webster Dictionary: A Critical Appraisal," in *Readings in Applied Linguistics* (3rd. ed.), ed. Harold B. Allen and Michael D. Linn. New York: Alfred A. Knopf, 1964. Reprinted in *About Language: A Reader for Writers*, pp. 127–39, ed. William H. Roberts and Gregoire Turgeon. Boston: Houghton Mifflin, 1986.

Soukhanov, Anne H., "Welcome to the Web of Words: The Lexicographer's Role in Observing and Recording the Changing Language," in *About Language: A Reader for Writers* (2nd ed.), pp. 121–33, ed. William H. Roberts and Gregoire Turgeon. Houghton Mifflin, 1989.

2. What do you think people generally see as the purpose of a dictionary? Ask five or six people. Are their responses all the same?

3. If your child said *I gived it to her,* would you offer a correction? What if your child said *I dreamed I was a clown?* Do you give different feedback on *gived* and *dreamed?* If you do, what does that tell you about how the next generation of English speakers will view these two past-tense verbs?

4. The verb *to be* is highly irregular, yet it isn't particularly susceptible to regularization. Why do you think this is so?

5. Occasionally a regular verb becomes irregular. For example, it is thought that *dived* preceded *dove* and *pleaded* preceded *pled.* What explanation can you give for these occasional irregularizations?

NOTES

1. *Webster's Third New International Dictionary* (unabridged) (Springfield, MA: Merriam-Webster, 1986), p. 4a.
2. Ibid.
3. *The American Heritage Dictionary of the English Language*, Third Edition. (Boston: Houghton Mifflin, 1992), p. vi.

3

NOUNS AND NOUN PHRASES

WHAT ARE NOUNS?

We will begin our discussion of English grammar with a close look at the lexical category **noun**. As with all other parts of speech, we will fold together our discussion of the two lowest levels of grammatical structure and discuss roots and affixes as part of our discussion of the word. Most of us recognize nouns by the traditional definition of their function: they name a person, place, thing, or idea. This is a reasonably useful definition, but it is not always sufficient to help us distinguish a noun from other parts of speech. It is additionally helpful to keep in mind that nouns have two inflectional markings: they are marked for **number** and **possession**.

When we talk about the *number* of a noun, we mean that it is either *singular* (one) or *plural* (more than one). Singular nouns in English have no special marking, but plural nouns are typically marked with the inflectional suffix *-s* (or *-es*): *pencils, jars, glasses*. We know, of course, that not every noun fits this pattern. There is a group of nouns that changes the vowel sound of the root to make the plural: *foot–feet, mouse–mice, woman–women*. There are other irregular plurals that do not fit any pattern, such as *oxen, children, deer*. All of these are holdovers from earlier forms of English, and now we learn them one by one. There is another important category of irregular plurals, the ones borrowed from other languages. Most of these are taken from Latin or Greek and tend to be more formal and less common than the Old English holdovers, such as *alumnus–alumni, criterion–criteria, phenomenon–phenomena, formula–formulae*. You are probably thinking that not everyone uses such singular and plural forms exactly the way we have described. There is a lot of evidence that people are trying to bring them more into the fold of the regular noun pattern. *Formulas* is fully standard and exists side-by-side with *formulae*. *Syllabuses* and *hippopotamuses* are already within the range of acceptability for most people. Others speak of *one criteria* and *one phenomena*. These are not considered to be standard English, but they are very common. If we say *one umbrella* and *one sofa*, why not *one criteria*? It is certainly likely that they will one day be considered the standard singulars and, when they are, *criterias* and *phenomenas* will probably follow. Meanwhile, if we want to stay within fully acceptable standard English, we need to overcome our instincts to think of them as singulars when we say them.

DISCUSSION EXERCISE 3.1

1. *Agenda* and *media* are historically plural forms, with singulars *agendum* and *medium*. What is the evidence that the plural forms have become accepted as singulars?

2. The plural of *fish* is historically *fish*, yet the regularized plural *fishes* has come into usage in recent times. Some people assign different meanings to the two plural forms. Do you know what those two meanings are?

3. The use of *alumnus–alumni* has one other interesting complication. They are derived from the Latin word meaning "student," and in Latin referred to male students. The corresponding female forms were *alumna* and *alumnae*. Would you object to naming the magazine for graduates of your college *The Alumnus*? That objection has been raised about the University of Michigan's *Michigan Alumnus*.

4. Why do you think the irregular plural *feet* has been more resistant to change than the irregular plural *syllabi*?

The other inflectional affix associated with nouns is the *possessive*. It also adds the suffix -*s*, separated from the noun root in writing with an apostrophe: *boy's, cat's, judge's*. Unlike the plural, the possessive form of nouns is completely regular. Even if the plural of the noun is irregular, its possessive fits the regular pattern: *men* for plural, but *man's* for possessive, for example. That is why we never hear any fluctuation in the use of the possessives and also why they are not very interesting as a subject of conversation. We do need to remember certain rules of spelling for possessives and also keep in mind that the possessive and the plural can occur together in one word. Although there is some variation from one handbook to another, the general spelling rule is that we add -'*s* to make a noun possessive, regardless of whether it is singular or plural: *car's, man's, men's, children's, Charles's*. But if the plural noun ends in -*s*, you simply add an apostrophe to make it possessive: *the Smiths' garage, the boys' uniforms*.

In addition to these telltale inflectional suffixes, nouns in English may also be marked by certain derivational suffixes—that is, suffixes that turn a root into a noun. Some common examples are *-er*, as in *dancer, singer, printer; -ment*, as in *government, filament, sediment* (notice that the noun root isn't always capable of standing on its own); *-ion*, as in *election.* Inflectional suffixes can occur together with the derivational ones and always appear at the end of the word: *dancers, dancer's, dancers'.*

DISCUSSION EXERCISE 3.2

1. Give some other derivational suffixes that turn roots into nouns.
2. Which of the following words are nouns? *visualize, national, realization, sincerity, fruity, engineer, dentist, happy, fearless, fearlessness, truthful, occurrence.*
3. Some noun roots can add derivational suffixes that do not change the part of speech. That is, the resulting word is still a noun, but with a somewhat altered meaning. What alteration in meaning is made by the suffix *-ette*, as in *kitchenette* and *cigarette*? What about the suffix *-ess*, as in *princess* and *actress*?

WHAT ARE SOME COMMON SUBCATEGORIES OF NOUNS?

We can use the various criteria we have mentioned above as guidelines for identifying a noun as distinct from some other part of speech, but we also know that they do not constitute an absolute definition that we can apply to any noun. Nouns fall into subcategories with their own special characteristics and do not all fit exactly the same mold. For example, we are familiar with the distinction between **common nouns** and **proper nouns**. *Common nouns* are written with lower-case letters and refer to general categories: *girl, teacher, ball. Proper nouns* are written with capital letters and designate a specific noun: *Mary, California, Fifth Avenue.* There are many differences in how these two subclasses of nouns behave, but one obvious one is that common nouns occur often in their plural forms, while the use of the plural for proper nouns is highly restricted. Other subclasses of nouns are **concrete nouns** and **abstract nouns**. Concrete nouns are the ones we can visualize: *table, chair, flag, hairdresser.* Abstract nouns are usually ideas or concepts with no clear visual image associated with them: *sincerity, construction, foolhardiness.* Again, the concrete nouns are the more typical nouns, in that they can be plural or possessive, and the abstract nouns are more limited in that respect. Nouns can also be divided into subclasses of **animate nouns** and **inanimate nouns**. Humans and animals fall into the first subcategory, whereas things fall into the second. Within the category of animate nouns, we further divide nouns into **human nouns** and **nonhu-**

man nouns. If a noun is human, we refer to it as *he* or *she*; if it is nonhuman or inanimate, we refer to it as *it*. If we hear a noise and think a human is making it, we ask *who is making that noise*? If we think it is nonhuman, or inanimate, we ask *what is making that noise*?

DISCUSSION EXERCISE 3.3

1. Suppose your neighbors come along with their brand-new baby wrapped in a yellow blanket. What difficulty might you have, grammatically speaking, finding out from them the baby's name or age?

2. Can you think of any circumstances in which people treat inanimate nouns as if they were grammatically human? Why do you think they do that?

3. How do you treat your family pet grammatically, as human or nonhuman? Do you differentiate grammatically between animals in your home and those in the zoo or the jungle? What about insects?

Other important grammatical subclasses of nouns are **count nouns** and **noncount nouns** (sometimes called **mass nouns**). Let's compare the noun *bean* to the noun *rice*. There are similarities in the things to which they refer: both are foods, and both occur in small, cylindrical pellets. Yet grammatically, we don't treat them alike at all. Suppose you want to count beans. One bean, two beans, three beans. . . . But if you want to count rice, you can't do it directly. You must provide some linguistic boundary for rice, like *grain* or *piece*. Then you can count one grain of rice, two grains of rice. . . . Or, you can put the rice in something and count that: one cup of rice, two cups of rice. . . . That is why we call *bean* a count noun and *rice* a noncount noun. What are the other differences between count and noncount nouns? (we use the convention * to indicate something that is generally considered to be "un-English.")

Count Nouns	Noncount Nouns
have plural forms: *beans*	do not have plural forms: **rices*
may not stand alone in the singular: **Bean is good for you*	may stand alone in the singular: *Rice is good for you*
can occur with *a* or *an*: *a bean*	cannot occur with *a* or *an*: **a rice*

Standard English also requires some very interesting differences in expressing quantities:

(too) many beans	*(too) much rice*
(too) few beans	*(too) little rice*
more beans	*more rice*
fewer beans	*less rice*

If we look at the patterns for expressing quantities, we can understand why people stray from the standard English pattern. We have two nouns that are not essentially different in meaning, yet standard English requires that we learn whether each is count or noncount and then make the appropriate grammatical distinctions. From the point of view of the speaker, this is an unnecessary complication of the grammar. We do not gain any meaning distinction; we just have to do more work. If you observe people's usage of count and noncount nouns, you will see attempts to avoid unnecessary work. Instead of distinguishing between *many* and *few*, people will say *a lot of beans, a lot of rice*. This is considered standard (as long as you spell *a lot* as two words) but informal. Or people might use *much* and *little* for both: *too much beans, too little beans*. Although these have not achieved standard acceptability, we can see the reason for their use: with no loss of meaning and no loss of a valuable distinction, people manage to make the overall system more predictable and less complicated, with *much* indicating a large quantity, and *little* indicating a small quantity. The situation is even more interesting when we are *comparing* quantities. Notice here that they are the same for the greater amount: *more beans, more rice*. But once again, for the lesser amount, we have to choose different words according to the rules of standard English: *fewer* for count nouns, *less* for noncount. What would be wrong with a simpler pattern that uses the same word for both, comparable to *more*? That is exactly what speakers of English seem to be asking every time someone says *less calories* or *less restrictions* or *less* any other count noun. How we treat the subclasses of count and noncount nouns is a very good example of how people collectively react to unnecessary burdens in their grammatical system. Without conscious agreement, there is movement toward a simpler, more regular pattern.

DISCUSSION EXERCISE 3.4

1. Which of the following nouns are count? Which are noncount? Use various grammatical tests to justify your decisions: *furniture, table, peace, student, sugar, university, greed.*

2. Some nouns in English can be both count and noncount, depending on how they are used in a sentence. *Beer* is an example of such a noun: *two beers, beer is a beverage.* Show how each of these nouns can be either count or noncount: *space, coffee, chocolate, time.*

3. You might have noticed that we used the word *amount* in the preceding paragraph to refer to both count and noncount nouns. If you are a grammatical purist, you might raise an objection to this usage. Traditionally, standard English has required that we speak of *amounts* of noncount nouns but *numbers* of count nouns: *the number of beans, the amount of rice.* What would you say is the status of this distinction? Is it nonstandard to use *amount* for both?

WHAT MAKES UP A NOUN PHRASE?

Nouns are the heads of larger groupings called **noun phrases**. Some of the modifiers of the noun are descriptive of the noun, like adjectives, prepositional phrases, and relative clauses. We will talk about these in later chapters. There are other modifiers of the noun that serve more to limit it, identify it, or place it appropriately in a conversation. These modifiers are part of what we call the **determiner** system. For example, the phrase in (1) below is a noun phrase:

    (1)   my first serious encounter with aliens

Encounter is the head noun. *Serious* is an adjective, and *with aliens* is a prepositional phrase, both describing *encounter*. *My* and *first*, on the other hand, are part of the determiner system. In example (2),

    (2)   the last person who saw him alive

person is the head noun, *who saw him alive* is a relative clause, and *the* and *last* are part of the determiner system.

    Noun phrases may take many different forms, so it is not easy to give absolute rules about the determiner system. Nevertheless, we can use the formula below as a guideline for talking about the determiner composition of noun phrases:

Noun Phrase = Predeterminer + Determiner + Postdeterminer + Noun

First, we must remember that the only required element in a noun phrase is the noun itself. Some nouns, like noncount or proper nouns, may be the only element of their noun phrase: <u>*Love*</u> *makes the world go 'round,* <u>*Michigan*</u> *is a beautiful state.* Some noun phrases may have only a determiner and a noun: <u>*The ground*</u> *is wet,* <u>*his sister*</u> *is visiting.* Some may have a determiner, a predeterminer, and a noun: *She is* <u>*such*</u> <u>*a liar*</u>*, I know* <u>*all*</u> <u>*the students*</u>. Some may have a determiner, a postdeterminer, and a noun: <u>*Her many friends*</u> *came to the party.* And others may have all four elements together: <u>*All John's many relatives*</u> *live in that house.*

Determiners

    Many noun phrases consist of just a determiner and a head noun. The most common determiners are **articles, demonstratives, possessive pronouns**, and **quantities**.

    There are two articles in English: the **definite article** and the **indefinite article**. The definite article is *the*. The indefinite article in standard English has two forms in the singular: *a* before a consonant sound and *an* before a vowel sound. *Some* can be considered the plural indefinite article, although it may also be considered a quantity. Some examples of noun phrases with articles are *the boy, a girl, an olive, some books.* Native speakers of English have no difficulty deciding how to choose between a definite and an indefinite article, but it is not easy to explain how we make that decision. Their traditional grammatical labels are misleading. For example, if I tell you *There is a book on the table. Please pick it up, "a book"* is just as "definite" as if I had said *Pick up the book on the table.* Perhaps a better way to look at the use of the articles is to imagine a speaker trying to communicate with a listener. If the speaker assumes that the listener already has a certain noun in mind, the speaker uses *the*. But if the speaker thinks the listener does not have the noun in mind, *a* (or *an*, or *some*) can be used to introduce the noun. Once it is placed in the mind of the listener, then the speaker can use *the*. I can say *the book* if I think you have already identified it in your mind; if not, I will introduce it by saying *a book* first.

DISCUSSION EXERCISE 3.5

1. Are *an honor* and *a use* exceptions to the rule governing the choice of indefinite article?

2. Why do you think there is fluctuation in standard English between *a* and *an* before a noun that begins with the sound *h* in an unstressed syllable: *a/an historical event, a/an hysterical patient, a/an hypothesis*?

3. Why do you think it works to start a conversation with a stranger by referring to "the president," "the sun," or "the moon"? How is that different from starting a conversation about "the secretary" or "the star"?

Demonstratives are like the definite article in function, with two differences: they also indicate the location of the noun relative to the speaker, and there is a cross-referencing rule that requires number agreement with the noun they modify. Two indicate that the noun is near the speaker, and two indicate that the noun is far from the speaker. You have probably figured out what they are:

>   this hat (singular, near the speaker)
>   these hats (plural, near the speaker)
>   that hat (singular, far from the speaker)
>   those hats (plural, far from the speaker)

Possessive pronouns make up another category of determiner: *my life, your idea, his ring, their reason. Quantities* are the fourth major kind of determiner: *many, several, enough, few, little, much, any, some, no, two,* for example. One other kind of determiner is a **possessive noun phrase**. You may wonder how we can use a noun phrase as part of a noun phrase, but that is typical of the nesting, hierarchical nature of language. Consider the noun phrases in (3) below:

(3)   The mechanic's advice
      My mechanic's advice
      That mechanic's advice
      Those mechanics' advice

You'll notice that, in each case, *advice* is the head noun of the noun phrase, but the determiner is also a noun phrase, with its own head noun *mechanic*'s. We might visualize this as:

[[the mechanic's]   advice]
         NP   NP

DISCUSSION EXERCISE 3.6

1. The use of demonstrative determiners is often accompanied by pointing to reinforce the idea of location. Some speakers also use the adverbs *here* and *there* for the same reason: *this here hat, that there hat*. This usage is nonstandard in English. Do you know of any languages in which such reinforcement of location is considered standard?

2. Make up noun phrases using all of the quantities listed in the paragraph above as determiners. What determines whether you use *few* or *little*? *many* or *much*? Does the same restriction hold for *enough*?

3. Think of sentences in which we use *any* as a determiner. Do you notice any restriction on its use? (Hint: we don't say **I have any books.*)

4. What is the head of each of the following noun phrases? What kind of determiner does each have?

your insecurity	a fool's mission
this explanation	that woman's child
the very important package	Sue's business

Predeterminers and Postdeterminers

As their names suggest, these occur surrounding the determiner, either before or after. Some common **predeterminers** are *all, half,* and *both*. These may be followed by the preposition *of*: *all (of) the people, half (of) the class, both (of) the students*. *What* and *such* are also considered predeterminers: *what a party, such a fool*, although their use is restricted to the indefinite article determiner. We can't say **what the party* or **such my fool*.

Postdeterminers express quantities as well and are called postdeterminers when they follow a determiner. Some common ones are cardinal numbers (*one, two, three* . . .), ordinal numbers (*first, second, third.* . .), and indefinite quantities like *several, many, few*. Some examples of noun phrases with postdeterminers are *my few friends, the first call, those many years, his six children.*

It should not trouble you that some words that express quantity can be used as predeterminers, determiners, or postdeterminers. It is often the case that a word's label is not inherent to the word, but rather it is derived from the way the word functions in a particular context. If you want to know what a word is in a noun phrase, look to see what else is in that noun phrase.

DISCUSSION EXERCISE 3.7
1. Label all the parts of the following noun phrases. Give all the information you can about each part.

    her bike   several questions
    the two children   both of his first choices
    all my sister's many friends   half that pie

2. Show how the word *many* can be used as a determiner or a postdeterminer. Show how *all* can be a predeterminer or a determiner.

WHAT ARE THE FUNCTIONS OF NOUN PHRASES?

Now that we know how to identify noun phrases and label their parts, the next step is to understand how they function in sentences. You can think of each sentence as a minidrama in which noun phrases play different roles. The most common roles, or functions, of noun phrases are **subject, direct object, indirect object, object of a preposition**, and **complement**. You will come to understand more about these functions as we discuss the other parts of the sentence, because they are primarily relational terms; that is, they describe how noun phrases interact with other parts of the sentence. Our goal at this point is to learn to identify them in sentences.

Subject

Although *subject* is a common grammatical term, and most of us have some intuition about what it is, it is surprisingly hard to define. When we refer to the *subject noun phrase* of a sentence, we often mean the doer of the action. In the sentences in (4), the subject noun phrase is underlined.

(4)   <u>Mary</u> left early.
      The <u>dog</u> jumped over the fence.
      My <u>children</u> caught the balloons.

But the doer of the action might *not* be the subject, as in (5),

> (5) The house was built by the contractor.
>
>   The exam was graded by the professor.

and often there is no "doer" expressed at all, as in (6).

> (6) This old house is a mess.
>
>   My many attempts at learning to play chess all failed.
>
>   The doctor's bill was astounding.

Probably a more reliable way for us to identify the subject noun phrase of a sentence is by its location. It is almost always the first noun phrase in the sentence and the one that immediately precedes the verb. By those criteria, you can identify the subject noun phrases in all of the above sentences.

We also need to keep in mind that, except for commands, all formal standard English sentences must have a subject, grammatically speaking. That is, there must be a noun phrase preceding the verb. For example, if we look out the window and see water falling from the sky, we must express this event by using a subject. Since there is no real subject, we use a "dummy" or "placeholder": *It's raining*. All speakers of English know that this subject is just a "dummy," so no one ever asks *What's raining?* Similarly, we use the word *there* as a dummy in sentences like <u>*There* are too many people on this bus</u>. Some grammarians call these placeholder words *expletives*.

DISCUSSION EXERCISE 3.8

1. Sometimes grammar books define the subject of the sentence as "what the sentence is about." What are the inadequacies of such a definition?

2. Remember that whatever is not the *subject* of the sentence is the *predicate*. Identify the subject and the predicate of each of the following sentences. (Remember too that subjects are noun phrases, not just nouns!)

    The playful child frightened the pony.

    My cat hid in the cupboard.

    Mrs. Waters just left.

    The telephone in the living room is portable.

    A small bird flew into the chimney.

    His computer is obsolete.

    The book that you just finished is a bestseller.

    Those three packages are for you.

    Both her parents attended her wedding.

    Love makes us happy.

3. Which of the following have placeholder subjects? How do you know? Can any be interpreted in two different ways?

   There is where I left my purse.

   It is snowing.

   There is an excuse for his tardiness.

   It is in the closet.

   It is too hot to eat.

Direct Object

Direct object eludes definition much the way *subject* does. It is often thought of as the "receiver of the action," which is helpful sometimes, as in the sentences of (7), where the underlined noun phrases are direct objects.

(7)   The girl hit <u>the ball</u>.
      The clown entertained <u>his children</u>.
      My aunt tossed <u>the salad</u>.

But again, the receiver of the action might *not* be the direct object, as in (8),

(8)   Mary received a blow.
      The ball was hit by the girl.

or the direct object might not be the receiver of the action, but rather comes into being as a result of the action, as in (9).

(9)   My friend wrote <u>a letter</u>.
      She invented <u>the wheel</u>.
      They built <u>a bridge</u>.

Although meaning criteria are sometimes helpful in identifying direct objects, once again location is probably a more reliable gauge of whether a noun phrase is a direct object. Direct objects usually come immediately after the verb. An additional important test for direct objects is the **passive test**. We will learn about passives in a later chapter, but for now you can see how the test works by comparing the two sentences in (10).

(10)   (a) The pitcher threw <u>the ball</u>.
       (b) <u>The ball</u> was thrown by the pitcher.

A noun phrase following a verb is likely to be a direct object, as in (a), if you can make a corresponding passive sentence just like (b) without changing the basic meaning.

DISCUSSION EXERCISE 3.9

1. Use meaning, location, and the passive test to identify the direct object noun phrase in each of the following sentences:

    Mary found Tom's keys.

    The engineer designed that building.

    He invited all my dearest friends.

    Everyone loved the charming little puppy.

    The clerk stamped the package.

2. Do both the following sentences have direct object noun phrases? What is your reasoning?

    Jane saw the president.

    Jane was the president.

Indirect Object

Indirect object is also not easy to define. Indirect objects almost always refer to the people who are, in some way, indirectly affected by the action. In the following sentence, *the teacher* is the indirect object.

(11)   Jimmy gave an apple to the teacher.

Jimmy is the subject, *an apple* is the direct object, and the person affected by this action is *the teacher*, the indirect object. Indirect objects are often preceded by the word *to* as in (11), or the word *for*, as in (12).

(12)   Rachel bought a sweater for Hannah.

In (12), *Hannah* is the indirect object. There is a very useful test for identifying the indirect object in a sentence, called **indirect object inversion**. You will notice in (13) that it is possible to rearrange the sentences of (11) and (12) by dropping the word *to* or *for* and moving the indirect object to the position immediately after the verb.

(13)   Jimmy gave the teacher an apple.

    Rachel bought Hannah a sweater.

In sentences like (13), we say that *the teacher* and *Hannah* are *inverted indirect objects*. We have not changed any of the meaning of the sentences, nor have we altered the relationships that hold among the noun phrases. Graphically, indirect-object inversion looks like this:

$$[\text{Verb} + \text{Direct Object} + \begin{Bmatrix} \text{for} \\ \text{to} \end{Bmatrix} + \text{Indirect Object}] \rightarrow [\text{Verb} + \text{Indirect Object} + \text{Direct Object}]$$

The formula is not foolproof and tends to work better with indirect objects preceded by *to* rather than *for*. Nevertheless, it is one more criterion to use in deciding whether to call a noun phrase an indirect object.

You need to be aware that the designation *indirect object* is controversial. Many linguists disagree with this traditional definition. For some, only the inverted version is called an indirect object, whereas the noun phrase following *to* or *for* is merely an object of a preposition (described in the next section).

DISCUSSION EXERCISE 3.10

1. Which of the following sentences have indirect objects? Which are inverted indirect objects?

   Keith gave Mary a present for her birthday.

   The professor taught those students linguistics.

   Kathy gave Sam a second chance.

   John cooked dinner for his folks.

   Jack read this book to Mildred.

   My daughter wrote this poem.

   The fabric is torn.

   Mary feels tired.

   The salesperson sold Maxine this radio.

   Sue sent flowers to Chuck.

2. How does indirect-object inversion affect our definition of direct object noun phrases? Are the noun phrases immediately following the verb always direct objects?

Object of a Preposition

Noun phrases that are *objects of prepositions* are easy to recognize if you know what a preposition is. We will have much more to say about prepositions in a later chapter, but for now we can say they are words that indicate the relationship of the following noun phrase to the rest of the sentence. Those relationships are many, including location, direction, accompaniment, and purpose. Prepositions link with a following noun phrase to form a constituent called a **prepositional phrase**. All the phrases in (14) are prepositional phrases, and the underlined noun phrases are objects of prepositions. Notice that once again we see a phrase nested inside another phrase.

(14)   in the barn (location)

       towards the fire (direction)

with <u>an</u> <u>escort</u> (accompaniment)

for <u>a</u> <u>good</u> <u>reason</u> (purpose)

As we mentioned earlier, you would not be wrong to call noun phrases following *to* and *for* objects of prepositions, as long as you also recognize that they have the special property of being able to move to another location in the sentence and drop their prepositions—that is, they have the capacity for inversion. Only indirect objects can do this, as evidenced by the ungrammaticality of sentences like those of (15).

(15)   Al studied algebra for a reason. → *Al studied a reason algebra.

Ted drove his car to the reunion. → *Ted drove the reunion his car.

Complement

Some noun phrases do not designate independent entities in a sentence. Rather, they serve to describe another noun phrase of that sentence. These noun phrases are called *complements*. Consider again the two sentences you were asked to compare in Discussion Exercise 3.9:

(16)   Jane saw the president.

Jane was the president.

You undoubtedly discovered that in the first one, *the president* is a direct object. It receives the action (to the extent that seeing is an action), and it works in the passive test: *The president was seen by Jane.* In the second sentence, *the president* fails the passive test and is not the receiver of an action. In fact, it is not a separate person at all, but a way of describing Jane. In this sentence, *the president* is called a complement, and since it describes the subject of the sentence, it is called a **subject complement**. (It may also be called a **predicate nominative** or a **predicate nominal**.) Now consider the underlined noun phrases in (17):

(17)   They considered the child <u>a genius</u>.

She declared her brother <u>a liar</u>.

In these sentences, the underlined noun phrase describes the direct object, and so it is called an **object complement**.

DISCUSSION EXERCISE 3.11

1. Pick out the objects of prepositions in the following sentences.

I'll meet you near the fence after school.

Larry cut the bread with a knife.

Put the box in the drawer under the sink.

She delivered the prescription to the pharmacy in the mall.

Ben went to the dance with Flo.

2. Identify the complements in the following sentences. Tell whether they are subject complements or object complements.

The earthquake was a frightening experience.

You are my best friend.

Everyone considered James an honest person.

The main dishes were beans and rice.

We declared Judy the designated driver.

(Note: Sometimes object complements sound better if you insert *to be* in front of them.)

3. Are the following sentences identical in structure?

The teacher taught Sam a lesson.

The teacher considered Sam a fool.

What is the grammatical function of each noun phrase in these sentences?

VERBAL NOUNS AND NOUN PHRASES

There is one kind of noun that deserves special attention, a **verbal noun**. You might not recognize verbal nouns as nouns at first because their meanings tend to be actions rather than things. These are words that are built from verbs, but they exhibit many of the properties of nouns. There are two kinds of verbal nouns: **gerunds** and **infinitives**.

Gerunds are verbs with the suffix -*ing*: *laughing, coughing, playing*, for example. What makes them nouns? One reason we call them nouns is that they can be heads of noun phrases, with many of the usual modifiers that occur with nouns. Look at the noun phrases of (18):

(18)   His laughing annoyed her.

All that coughing is disturbing the musicians.

The child's crying made me nervous.

In these, we see that gerunds can occur with determiners such as possessive pronouns, demonstratives, and possessive noun phrases, as well as with predeterminers. Verbal nouns are abstract nouns, so the determiner system is somewhat limited, as is the case with all abstract nouns. Nevertheless, we can see that the gerunds in the above examples are clearly the heads of noun phrases. Noun phrases with gerunds as their heads are called **gerundive phrases**. Gerundive phrases may be further expanded with modifiers, like manner, place, or time, as in the sentences of (19).

(19) His laughing like a hyena annoyed her.
All that coughing in the audience is disturbing the musicians.
The child's crying all night made me nervous.

Gerundive phrases also perform the typical grammatical functions of noun phrases. They may be the subject of the sentence, as we see in the sentences (19). They can be direct objects, as in the sentences of (20).

(20) She resents his laughing like a hyena.
The musicians don't like all that coughing in the audience.
I couldn't bear the child's crying.

They can be objects of prepositions, as in the sentences of (21).

(21) She teases him about his laughing like a hyena.
The musicians are upset over all that coughing in the audience.
I was disturbed by the child's crying.

Gerundive phrases can also be complements:

(22) His most annoying trait is his laughing like a hyena.
The cause of the noise is all that coughing in the audience.
The reason for my irritation was the child's crying.

So we see that gerundive phrases behave more or less the way other noun phrases do.

The other type of verbal noun is called an **infinitive**. Infinitives are verbs with the word *to* in front of them: *to talk, to love, to run,* for example. These may serve as the heads of noun phrases called **infinitival phrases**. These too exhibit many of the properties of noun phrases. Although they are not as versatile as gerundive phrases, they may be subjects, direct objects, and complements, as can be seen in the sentences of (23).

(23) subject: To give up now would be foolish
direct object: Everyone desires to live in peace
complement: His first instinct was to run away

DISCUSSION EXERCISE 3.12

1. You probably noticed that we didn't give any examples of gerundive and infinitival phrases as indirect objects. Why do you think they can't be indirect objects?

2. Find the gerundive phrase in each of the following sentences. Tell what its grammatical function is in the sentence.

   You learn by studying every day.

   Hiking in the woods is fun.

   We don't mind leaving early.

   Thank you for not smoking in my car.

   The most exciting activity at camp is swimming in the lake.

3. Find the infinitival phrase in each of the following sentences. Tell what its grammatical function is in the sentence.

   The important thing is to print your name legibly.

   She prefers to work alone.

   His goal in life is to make a lot of money.

   To bail out now would be a mistake.

   Rose liked to bask in the sun.

We have now examined the roles of all the players in the sentence minidrama. In the next chapter we examine what they are doing or what is being said about them. The constituents that express this part of the drama are verbs and verb phrases.

REFLECTIONS

1. There seems to be an increasing use of the apostrophe in the plural. You might see an apartment building that advertises *"studio's for rent"* or a fast food restaurant that advertises *"taco's"* and *"burrito's."* What might be the explanation for this widespread deviation from conventional spelling?

2. *Data* is another example of a noun that is historically plural. Its Latin singular is *datum*. *Data* is still used in English as a plural noun, but it also occurs as a singular noncount noun. What would be an example of this usage? What does your dictionary say about the status of *data* as a singular noun?

3. Some language change is motivated by political or social change. For example, many women have objected to the use of the feminine suffix *-ess* as an add-on to an occupation, and prefer to be called a *poet*, not a *poetess*; an *author*, not an *authoress*. Can you think of any other nouns with *-ess* that have fallen into relative disuse?

4. A related change is the replacement of occupations that have traditionally ended in the suffix *-man* with gender-neutral terms: *postal carrier, fire fighter*. Can you think of others?

5. Visit your local supermarket and see how many product labels advertise products as having *"less calories."* At this point, could we consider *less calories* standard English?

6. Some people do not use *an* in casual speech. Instead, when the noun begins with a vowel sound they insert a sound called a *glottal stop*, which sounds like a catch in the throat. Listen for it the next time you hear someone say *"a apple"* or *"a orange."*

7. Notice that if we do indirect-object inversion on a sentence, the indirect object ends up in the position where we expect to find the direct object, right after the verb:

   Kathy gave Sam a second chance.

   That means we can't say that the noun phrase after the verb is *always* the direct object. How does indirect-object inversion affect the passive test? Is it a totally reliable test for the direct object?

8. One common nonstandard usage appears in the following sentence:

   We appreciated him buying the tickets.

   In its most ordinary interpretation, this sentence is nonstandard. What makes it so?

9. Another rule passed down from the eighteenth century is a prohibition against splitting an infinitive. This means nothing should come between *to* and the verb. *To never see daylight* is an example of a split infinitive. What do you think the status of this rule is in modern English usage? Consult several style manuals to see if there is consensus among them for formal, written English.

PRACTICE EXERCISES   (Answers on p. 257)   _____

1. Underline all the nouns in the following sentences. Name the subcategories to which they belong.

   Example: *Mike bought some candy for his girlfriend.*
   Mike: proper, human, concrete
   candy: common, inanimate, noncount, concrete
   girlfriend: common, human, count, concrete

   1. Calcium is an element necessary for strong bones.

   2. The instructor asked the students to review the exercises.

   3. A worker in the factory sensed the dissatisfaction among his colleagues.

    4. Cats and dogs provide friendship and love for lonely people.

    5. The state animal of Minnesota is the gopher.

2. Identify all the nouns in the following sentences and tell whether they are count or noncount nouns.

    1. The wind tore a hole in the tent.

    2. Too much sugar in your diet leads to poor health.

    3. My mother enjoys wine with her dinner.

    4. Dinner is the most important meal of the day.

    5. The babysitter fed the baby milk with his peas.

3. Label all the predeterminers, determiners, postdeterminers, and heads of the following noun phrases.

    1. both my older sisters

    2. half of his geometry class

    3. what an exciting event

    4. the girl's third attempt

    5. three boats

    6. my seven cousins

    7. such a shame

    8. all the children

4. What specific kind of determiner (definite article, possessive pronoun, etc.) appears in each of the noun phrases above?

5. Identify all the noun phrases in the following sentences. Tell the grammatical function of each noun phrase you identify.

    1. The weekend is my favorite part of the week.

    2. Martha's boss gave her a bonus.

    3. Bob considered the doctor the answer to his prayers.

    4. A quaint old boat drifted down the river near Sally's farm.

    5. The cashier gave the receipt to the woman in the fur coat.

6. Give the noun phrases in these sentences that play the indicated grammatical roles:

    1. The child gave a present to her mother on Mother's Day.

        subject: _____

        direct object: _____

        indirect object: _____

        object of a preposition: _____

    2. The first day of April is my favorite time of the year.

        subject: _____

        subject complement: _____

        object of a preposition: _____

        object of a preposition: _____

7. Which of the following sentences have indirect objects? Which of the indirect objects are inverted?

    1. The grateful student sent the teacher a note.

    2. My father cooked dinner for the whole family.

    3. Randi bought a toy for her child for Christmas.

    4. Pam read Jim the instructions.

    5. The trainer fed the lions raw meat.

8. Label the roles of all the noun phrases in the following sentences.

    1. Mr. Allen taught the boy geometry.

    2. Mr. Allen considered the boy an idiot.

9. Underline the verbal noun phrase in each of the following sentences. Is it a gerundive or an infinitival phrase? What is its grammatical function in the sentence?

    1. Winning a race is exhilarating.

2. You achieve success by working hard.

3. I've always liked to hike in the woods.

4. To live honestly is my main goal in life.

5. You can't accuse him of not trying.

6. The most important thing is to remain calm.

7. I can't tolerate all this whining.

8. She appreciated learning a new skill.

9. The hardest part is waiting for the results.

10. They expect to arrive tomorrow.

10. Which noun phrases in the following sentence are nested inside another phrase?

The young woman's father purchased the farm at the foot of the hill.

4

VERBS AND VERB PHRASES

WHAT ARE VERBS?

Verbs are often considered to be that lexical category that indicates the action of the sentence, but you can probably imagine some of the shortcomings of this definition. We know that sometimes actions are expressed by nouns, as in

(1) The appointment of a new dean

Similarly, we know that not all verbs express actions. They may express, among other things, a sense, as in (2), a perception of another person (3), a mental state (4), or merely serve a connecting function (5).

(2) Lora <u>feels</u> sad.
(3) Lani <u>seems</u> contented.
(4) Jim <u>expects</u> a package in the mail.
(5) Steve <u>is</u> the youngest member of the club.

Sometimes we can recognize verbs by their derivational suffixes. There are several such derivational suffixes—that is, suffixes that turn a root into a verb. Some common ones are *-ize (nationalize, subsidize), -ate (educate, motivate), -ify (solidify, verify)*.

DISCUSSION EXERCISE 4.1

1. Give some other verbs formed by adding the suffixes listed above.

2. *-esce* is a less common derivational suffix for verbs. How many verbs can you think of that end in it?

But verbs often do not have any derivational suffix and are better recognized by the various forms they may take. For example, we have already talked about the infinitive of the verb. Verbs may appear with the word *to* in front of them; this combination is called *the infinitive of the verb*. As we saw in the last chapter, *infinitives* are usually verbal nouns and serve the function of a noun in a sentence, but they are formed from verbs and are a convenient way to refer to the verb when you want to talk about it ("Let's talk about the verb *to run.*"). You may also see the verb used without *to* in front, sometimes referred to as the **bare infinitive** or the **base form**. ("Let's talk about the verb *run.*"). For its part, the bare infinitive is commonly used with what is known as a helping verb, as in the sentences below.

(6)   He can <u>type</u> fifty words a minute.

You may not <u>leave</u> yet.

Another form that all verbs take is the **present participle**, which is formed by adding the suffix *-ing* to the base form: *laughing, singing, being, feeling.* The present participle and the gerund have the same form but different functions. The gerund, you will remember, is a verbal noun and plays the role of a noun in sentences. The present participle is always part of a verb and occurs with a helping verb. Compare sentences (7) and (8).

(7)   Talking relieves tension.

(8)   I am talking to you.

In (7), *talking* is a gerund and serves as the subject of the sentence. We could modify it with a determiner (*his talking*), or with an adjective (*his incessant talking*), for example. But in (8), *talking* is part of the verb: it does not play any noun functions, nor can we modify it with noun modifiers: **I am his talking to you; *I am his incessant talking to you.* In (8), *talking* is a present participle. We follow traditional grammar in using this term, but it is misleading because it is not present; in fact, it carries no time by itself. Notice how the time of the action changes in each of the sentences below while the present participle stays the same.

(9)   I am studying. (now)

I was studying. (yesterday)

I will be studying. (next weekend)

A third form that verbs take is the **past participle**. Also misnamed because it carries no time by itself, the *past participle* is the form of the verb

that occurs in the blank space after *have* —————— or *had* —————— .
Often the past participle is formed by adding the suffix *-ed* to the base form,
as in *talked, laughed, danced*. In other cases, we add *-en* to the base form:
eaten, beaten, taken. But for a very large number of verbs, the form of the
past participle is unpredictable and must be learned on a case-by-case basis. Sometimes the vowel of the base form changes, as in *written, spoken, dri-*
ven. Sometimes the vowel changes but no suffix is added: *run, sung, drunk*.
Sometimes the past participle seems unrelated in form to the base form, as
in *I have bought, gone, taught*. We will have more to say about the irregularity of past participles and its effect on people's usage. For now, our goal is
to recognize the past participle in a sentence and realize that, just like the
present participle, it carries no time of its own. The sentences below illustrate how the time frame may change even though the past participle remains the same.

(10) She has often studied hard. (in the recent past)

She had often studied hard. (before something else happened
in the past)

She will have studied hard. (by some time in the future)

Because the infinitive, the present participle, and the past participle carry
no time of their own, they are known as **nonfinite verb forms**.

DISCUSSION EXERCISE 4.2

1. Give the present and past participles of each of these verbs: *walk, run, do, catch,*
fight.
2. Identify the nonfinite verb forms in the following sentences:

Are you speaking to me?

He needs to see a doctor.

Sam has already locked the door.

Must you make such a mess?

Why hasn't she consulted a specialist?

They aren't listening to the lecture.
3. Which *-ing* words are gerunds and which are present participles? How do you
know?

Spelling has always come easy to me.

I heard that you were leaving.

She is expecting us in an hour.

You learn by studying.

I heard laughing in the audience.

Verbs also have two forms that do carry time—the present and the past. Typically, verbs will add the suffix -*s* or -*es* to the base form to mark the present tense, but only if the subject is a singular noun phrase or the pronoun *he, she,* or *it*. These are known as *third-person-singular subjects.* You can see that the suffix appears when a third-person-singular subject is used, as in the sentences of (11), but with any other subject the present tense form looks just like the base form, as in (12).

(11)   This child look**s** hungry.
       She smile**s** too much.
       It need**s** to be proofread.

(12)   These children look hungry.
       You smile too much.
       They need to be proofread.

The past tense is typically marked by adding the suffix -*ed* to the base form. Unlike the present tense suffix, it is used for all subjects and not restricted to third-person singular.

(13)   The children look**ed** hungry.
       I sav**ed** the whale.
       We laugh**ed** about the misunderstanding.

The past tense, like the past participle, is subject to many irregularities. There is no pattern to the irregularities; they are remnants of older patterns in English and now must be learned one by one. Like the past participle, sometimes the past tense is marked by a vowel change, as in *ran* and *sang.* In other instances the past tense seems to bear no relation to the base form, as in *went* and *bought.*

We will talk more about verb irregularities in the next section. Our purpose here is to lay out the different forms in which any one verb can appear. We have already identified the three nonfinite forms: the infinitive, the present participle, and the past participle. The two other forms, the present tense and the past tense, because they are identified with a particular time, are known as **finite verb forms**. If we can recognize and identify these five forms of a verb, we have all the basic tools we need for talking about verbs and how people use them.

DISCUSSION EXERCISE 4.3

1. For each bare infinitive, give the corresponding present participle, past participle, third-person-singular present tense, and past tense: *talk, laugh, dance, educate, synthesize, qualify.*

2. Which of the following words are verbs and which aren't? How do you know? *special, character, invent, compute, mechanic, sadly, refer, equate.*

3. In many cases, a word can be either a noun or a verb. Can you demonstrate this for *smell, pit, laugh, part?*

4. In other cases, there are two words that differ only by the placement of stress, one a noun and the other a verb. *Subject* is one example. Can you think of others? Does the word *influence* fall into this pattern for you?

5. English has had a tendency over the years to create verbs out of nouns, a process known as *conversion*. *Market* and *audition* are two such examples. Can *eyeball* be a verb? What about *input* and *output?* Can *disrespect* be a verb?

WHAT ABOUT THE EXCEPTIONS?

Verbs present problems for speakers of English because the patterns are not reliable, as we have already seen. There are no problems associated with the infinitive and the present participle. These are always regular: the infinitive is always *to* + the base form, and the present participle is always the base form + *-ing*. The third-person-singular present tense has a few irregularities, but not enough to cause major concern. The real problem for us is the past tense and the past participle. These are the forms that have many exceptions and require us to memorize them rather than apply a rule.

Let's look again at the regular pattern for verbs. The verb *to laugh* is a good illustration:

laugh (base form)

laughed (I laughed all day yesterday.) (past tense)

laughed (I have never laughed so much.) (past participle)

In the regular pattern, the past tense and the past participle both add *-ed* to the base form. Whenever a new verb is added to English, it follows this pattern:

eyeball (base form)

eyeballed (I eyeballed the new contract.) (past tense)

eyeballed (I have often eyeballed new contracts.) (past participle)

But this pattern was not always the dominant one for English. Old English had a much more complicated way of forming the past tense and the past participle, somewhat unpredictable even then, and many of our current forms are holdovers from long ago. As a consequence, we can never be sure whether the past tense or the past participle will use the suffix *-ed*, nor can we be sure that the past tense and the past participle will be the same form. Here are some combinations that occur frequently:

Past Tense and Past Participle Are the Same, but Do Not Add *-ed*

Base Form	Past Tense	Past Participle
catch	caught	caught
hit	hit	hit
buy	bought	bought
bring	brought	brought
have	had	had
keep	kept	kept
sleep	slept	slept
swing	swung	swung

Past Tense and Past Participle Both Change the Vowel of the Base Form, but Differ from Each Other

sing	sang	sung
drink	drank	drunk
swim	swam	swum

Past Tense Changes the Vowel and the Past Participle Adds the Suffix *-en* or *-n* to the Base Form

take	took	taken
see	saw	seen
fall	fell	fallen
know	knew	known

Past Tense Changes the Vowel and the Past Participle Shortens the Vowel of the Base Form and Adds the Suffix *-en*

write	wrote	written
drive	drove	driven
ride	rode	ridden
strike	struck	stricken

Past Tense Changes the Vowel and the Past Participle Adds *-en* to the Past Tense Form

speak	spoke	spoken
break	broke	broken

Still other combinations don't lend themselves to easy description:

do	did	done
go	went	gone
be	was, were	been

It is easy to see why speakers of English have trouble remembering the past tense and the past participles of irregular verbs. One common non-standard usage results from assuming that the past tense and the past participle are the same, just as they are in the regular pattern. So you might hear people say such nonstandard sentences as

(14)   I seen it before.

   We done it already.

These assume that the past participle also serves as the past tense. In other cases, people assume that the past tense also serves as the past participle and say sentences like

(15)   He has ran a mile.

   She has went for milk.

DISCUSSION EXERCISE 4.4

1. Although people may be judged uneducated for using sentences like those in (14) and (15), their meaning is understood. Why do you think people get the intended meaning despite the confusion of past tense and past participle?
2. Because confusion of past tense and past participle carries with it some social stigma, educated people are sensitive about their usage. Given the range of unpredictability of these forms, we all fall prey to anxiety about some verbs. Can you think of any that you are unsure of? Did any of the ones listed above surprise you?
3. If you can't think of any, you might consider the verb *sneak*, or *lie* and *lay*. Does everyone agree on the past tense and past participle forms of these verbs?

Given the variability of irregular past tense and past participle forms, you can see that they would be unstable in the language. Language users, as you know, prefer simple, regular patterns to follow. We certainly do not lose any ability to communicate if we make an irregular verb fit the regular pattern, and this is precisely what speakers of English have been doing for centuries. There used to be many more irregular verbs in the language than there are now. Gradually, the irregular ones get replaced by regular ones. The most common ones are heard so frequently that people tend to retain them in standard English, but less common irregular verbs are highly vulnerable to regularization. We take for granted now that *climb, walk, drag, help, ache, laugh,* and *yield* are regular verbs, but they are among many that became regular over time. This process continues into modern English.

When a regular form is replacing an irregular form, both are used for a period of time. Here again is fertile ground for linguistic anxiety. Not only are we uncomfortable with two ways of saying the same thing, but we also know that regularizations are considered uneducated when they first appear. (If you doubt this, think of your reaction to someone who says *I knowed it.*) So if we hear two different past tense forms or two different past participles for the same verb, we want to know which is "correct." But in the absence of a unique authority, we have to make judgments about whether a new form is considered standard or whether an older form is considered old-fashioned. At a certain point in the transition from old to new, the two forms coexist and no such judgments can be made. Judgments are clearer before and after this point. No one wonders any more whether the past tense of *laugh* is *low* or *laughed*. But no such clear judgment can be made about *dream*, for example. Is it more correct to say *I dreamt* than *I dreamed*? There is nothing about people's usage or people's reactions to their usage that would lead us to choose one over the other.

DISCUSSION EXERCISE 4.5

1. Are the following verbs regular or irregular in their past tense and past participle forms? *plead, stride, strive, sweep, dive, creep, shine, leap, kneel, slay, light.*

2. If you find that there are competing forms, what would help you decide if only one or both were considered standard?

3. Consider the verbs *lie* (as in *lie down*) and *lay* (as in *lay an egg*). They are undoubtedly in a period of transition. These are the forms that have been considered standard until now:

lie:	*lay* (past tense)	*lain* (past participle)
lay:	*laid* (past tense)	*laid* (past participle)

    What evidence is there that people don't always use these verbs as described above? What is your own assessment of the status of this alternate usage?

WHAT ARE SOME COMMON SUBCATEGORIES OF VERBS?

The characteristics of verbs that we have described apply generally to verbs but, just as is the case for nouns, there are different types of verbs and some exhibit more of the typical verb properties than others. In our discussion so far, the kind of verb we had in mind was a **main verb**. Main verbs are what we usually think of as verbs: they express actions or states of being, they have all of the five forms we described, and they can occur alone, independent of any other verb. The underlined word in each of the sentences below is a main verb.

(16) Mike <u>studies</u> statistics.

My aunt <u>collapsed</u> in the hallway.

Cary <u>looks</u> healthy.

There is another subcategory of verbs called **helping verbs**. These, as their name suggests, are used to support a main verb and do not occur by themselves in sentences. Helping verbs themselves come in two types: **auxiliary verbs** and **modal verbs** (also called **modal auxiliaries**).

There are three auxiliary verbs in English: *be, do,* and *have*. They occur together with a main verb in sentences, and when an auxiliary verb is present, the main verb occurs in one of its nonfinite forms: the present participle, the past participle, or the bare infinitive. The sentences below will give you an idea of how auxiliary verbs function in sentences.

(17) I <u>am</u> waiting for an answer.

<u>Do</u> you think it will rain?

<u>Have</u> you seen the report?

The car <u>was</u> washed by the students.

You will notice that the auxiliary verbs carry no significant meaning of their own; it is the main verb that carries the meaning of the action or the state. The main job of the auxiliary verb in these sentences is to indicate the time of the action, since the nonfinite verb forms cannot do this by themselves.

DISCUSSION EXERCISE 4.6

1. Show how you can change the time of the action in the sentences of (17) by changing the time of the auxiliary verb.

2. Sometimes in conversations auxiliary verbs will occur in sentences without a main verb, as in *Yes I am* or *He did*. Does this contradict our claim that auxiliary verbs must occur with a main verb?

3. Although *be, do,* and *have* are the three auxiliary verbs of English, they may also function as main verbs. What makes the underlined verbs below main verbs, not auxiliaries?

    Rula <u>did</u> her homework every night.
    The cats <u>are</u> playful.
    I <u>have</u> enough money to take a trip.

As speakers of English, you are also aware that the auxiliary verbs all have certain irregularities. Because we use them so much, we have not weeded out their irregularities. *Do* is irregular in the past tense (*did*), the past participle (*done*) in the third-person-singular present tense (*does*). *Have* is irregular in the past tense (*had*), the past participle (*had*), and the third-person-singular present tense (*has*). (*Have* used to be more regular, but the pronunciation of the sound *v* sometimes got lost in the middle of words). The verb *be* is the most irregular of all verbs and requires a closer look.

The first thing we notice about the verb *to be* is that it has more different finite forms than the other verbs. We list all its forms below.

to be (infinitive)
being (present participle)
been (past participle)
am, is, are (present tense)
was, were (past tense)

You'll notice that in the present tense there is a choice of three different
forms, while all other verbs have only two. In the past tense there is a
choice of two forms, while all other verbs in the language have only one.

DISCUSSION EXERCISE 4.7

1. Because *to be* is so irregular, people may try to make it more regular in its usage.
   One common nonstandard usage occurs in sentences like *we was so happy,
   they was laughing, you was lying to me*. How can these be explained as an attempt to make *be* more like other verbs?
2. Do you ever hear attempts to simplify the present tense of the verb *to be*? What
   would be some examples of this?

The other kind of helping verb is a *modal*. There are nine modal verbs
in English: *will, would, shall, should, can, could, may, might, must*. (*ought to* is
sometimes added to this list.) They each have only one form: there is no **to
shall* or **musting* or **woulded*, for example. Like the auxiliaries, they always
occur with a main verb:

> (18)   Jan must leave now.
>   Tom might be absent.
>   Fran should tell him.

But unlike the auxiliaries, they do carry some meaning of their own. In fact
they carry a wide range of different meanings and nuances of meaning that
we learn as we learn English but are very hard to spell out in exact and predictable terms. Consider the meanings conveyed by the modal in each of
the following:

> (19)   Chuck will sell his house. (future certainty)
>   As a child, Irene would hide in the garden. (repeated past
>   activity)
>   The toddler might hurt himself. (possibility)

> The child may eat now. (possibility or permission)
> I should call her. (obligation)
> He can swim a mile. (ability)
> They should arrive by seven. (probability)

You will notice that some may carry more than one meaning, so spoken in isolation the sentence might mean more than one thing, such as *the child may eat now*. We usually know from the context which meaning was intended.

DISCUSSION EXERCISE 4.8

1. What is the meaning of the modal in each of the sentences of (18)?
2. Which modal carries the meaning of necessity? Past ability? Advisability? Give some examples of sentences that demonstrate these meanings.
3. What meanings are conveyed by *could* in the sentence *She could swim ten miles*? (Hint: expand the sentence to put it in a larger context.)
4. One of the few grammar rules explicitly handed down from one generation to the next is the rule about *can* and *may*: *can* is for ability, *may* is for permission. To what extent do you think this rule is in effect in modern English usage?
5. There is an obsolete grammar rule about *will* versus *shall* that is not in effect in modern American English. Do you know what it is? The answer appears in Reflections 2 at the end of the chapter.

There are two other important uses for modal verbs. They occur in the second part of what are called *hypothetical if-then* statements (although *then* itself is optional):

> (20) If you had the money, (then) you could go to Europe.
>
> If Dan applied himself, (then) he might get good grades.
>
> If Carla smiled more, (then) she would seem kinder.

The *if* part of the statement says something that is contrary-to-fact (you don't have the money, Dan is not applying himself, and Carla doesn't smile more); the *then* part expresses a condition that would result if it were a fact. We say that these modals function as **conditionals**.

Another use for some modals is to soften commands and make them seem less blunt or rude. Here they tend to lose their individual meanings and are more or less interchangeable:

> (21) Could you help me?
>
> Can you help me?

Might you help me?

Would you help me?

You will notice that each of these has a literal meaning and can be answered literally:

(22)   Could you help me? I could yesterday, but I can't today.

I could if I had the time.

Can you help me? Yes, I am physically capable of helping you.

Might you help me? I might if I saw something in it for me.

Would you help me? I would if I were a more generous
person.

But there is another use for these questions that requires no verbal response at all. Opening a door or relieving someone of a heavy package might be a sufficient response if you interpret them as commands for assistance (albeit softened and polite) rather than requests for information.

DISCUSSION EXERCISE 4.9

1. Soften the following commands by using a modal:

Get me a beer!

Stop talking!

Lend me five dollars!

2. Explain how these statements could mean different things in different contexts:

You might change your attitude.

You may want to check your spelling.

3. Sometimes modals are used as conditionals with no accompanying *if*-statement, but there is still an implied *if*-statement. Supply an appropriate *if*-statement for the following conditionals:

You could get better grades.

I would not do that.

He would do anything for her.

WHAT IS VERB TENSE?

Verbs, as we already know, are those words that are capable of carrying the time of the action. We are used to thinking of the **time** of the verb as the **tense** of the verb, but they are not exactly the same thing. Let's talk about *time* first. English has three different times that we associate with verbs: present, past and future. The present, despite its name, does not refer to activities going on at the moment. Rather, it refers to general facts and activities that include right now but cover a much wider range of time. The examples below will give you some idea of the present time:

(23)   My dog loves pickles.
        Jane reads historical novels.
        The workers punch time cards.
        This car needs a brake job.
        The Bronx is part of New York.

The past refers to an action or fact prior to the time of utterance:

(24)   I studied French in Paris.
        The enemy invaded at dawn.
        It snowed yesterday.
        They bought a new house.
        Betty left in a huff.

The future, of course, refers to an action or fact subsequent to the time it is said or written:

(25)   You will regret this.
        The police officer will file a report.
        The world will end next Friday.
        My aunt will visit us in June.

These three times make up what we call the **simple tenses**. So, if you thought time and tense were the same thing, you were right up until this point. We can talk about the times we described above as the **simple present**, the **simple past**, and the **simple future**.

DISCUSSION EXERCISE 4.10

1. Given what we already know about verbs, it is easy for us to describe how each of these tenses is formed. We know that if the subject of the present tense verb is a singular noun phrase or *he, she, it*, we must add the suffix *-s*. What form does the verb take otherwise?

2. The past, in its regular formation, adds the suffix *-ed* to the base form. Does it matter what the subject is? Give some examples of regular past tense verbs. What are some exceptions?

3. The future tense is expressed with the modal *will*. What form of the verb follows *will*? Does it matter what the subject is? Are there any exceptions?

The simple tenses allow us to talk about actions occurring at a variety of times relative to when we say them, but they are limited in what they can express about the time of an action. For example, as we saw above, they do not let us talk about something happening right now. If I say *Jane reads historical novels*, it doesn't tell me what she is doing right now. Similarly, if I wanted to describe an event as background to another event, I couldn't do it with the simple tenses. Nor could I distinguish something very far in the past from something in the recent past. To give us a richer and more complete way of referring to the time of actions, English provides us with what is called **aspect**. Aspect never occurs on its own; rather, it combines with time to form the **complex tenses**. Thus, English has three times, which may occur on their own to express a simple tense, or they may combine with aspect to form the complex tenses:

Time (Present, Past, Future) = Simple Tense
Time + Aspect = Complex Tense

To understand what this means, we need to understand what *aspect* is, and that is best accomplished by illustration.

There are two different aspects in English, the **progressive aspect** and the **perfect aspect**. The *progressive aspect* allows us to describe actions as backgrounds to other actions, or actions in progress. The following examples will show you how the progressive aspect works.

(26)   As I speak, Mary <u>is writing</u> her letter of resignation.
When Pat arrived home, the children <u>were playing</u> quietly.

Although the progressive may occur without another action overtly expressed, another "anchor" action is always implied:

(27)   My roommate was snoring (when I passed his room).
The kitten is hiding (as I say this).

You will notice that the progressive aspect involves the use of the present participle of the verb, but the present participle cannot occur as a verb all by itself. We wouldn't say *My roommate snoring or *the kitten hiding. The present participle is a nonfinite verb form, meaning it cannot carry time. So what we do is use it together with the auxiliary verb *be* to express the progressive aspect. When a time is attached to the verb *be*, the progressive tenses are formed, as indicated below:

Progressive Tenses = (Be + Time) + Present Participle

Now, since there are three different times, there are three different progressive tenses. The tense will change as you change the time of the verb *be*, as you can see in the following set of sentences.

(28) She <u>is</u> waiting. (present progressive tense)

She <u>was</u> waiting. (past progressive tense)

She <u>will be</u> waiting. (future progressive tense)

DISCUSSION EXERCISE 4.11

1. Use the frame *The boy study Spanish* and give the sentence in the simple present, the simple past, the simple future, the present progressive, the past progressive, and the future progressive.

2. Change the subject to *the boys* and do the same thing. What changes do you have to make?

3. Tell the tense of the verb in each of the following sentences:

    The house lacks character.

    Polly was asking for a cracker.

    The storm will pass in an hour.

    We were expecting trouble.

    My uncle will be staying with us.

    The chef is preparing a banquet.

    Lightning destroyed the barn.

4. Sometimes one tense can express a time ordinarily reserved for another tense. For example, *I leave tomorrow* uses the present tense but actually expresses a future time. What tense is used in each of the following? Can it express a time outside its normal function?

    Paul is studying French at school.

    Sarah feels ill right now.

    The Germanic tribes invade England in AD 449 and drive out the Celts.

    So then he says "Leave me alone."

    The last two examples illustrate what is sometimes referred to as the *historical present.*

The second aspect of English is the *perfect aspect,* which serves to associate an action with a later time. Consider the following sentences, all of which contain the perfect aspect:

(29) I <u>have seen</u> that movie already.

The woman <u>had left</u> by the time her sister arrived.

By our next class, you <u>will have read</u> the chapter.

In the first, the action is in the past, but the use of the perfect aspect connects it to the present, a later time. That is, the sentence suggests that you saw the movie recently, not in the very far past. In the second, both actions are in the past, but the woman's leaving precedes the arrival of her sister. In the third, they are both future events, but reading the chapter comes first and is connected to a later event, the next class. To express the perfect aspect, we use the auxiliary verb *have* and the past participle of the main verb. When we attach a time to *have*, we form the perfect tenses, as illustrated below.

Perfect Tenses = (Have + Time) + Past Participle

This gives rise to three more complex tenses:

(30)   She <u>has waited</u>. (present perfect tense)
       She <u>had waited</u>. (past perfect tense)
       She <u>will have waited</u>. (future perfect tense)

DISCUSSION EXERCISE 4.12

1. The use of the perfect tenses out of context seems odd because they imply some connection to a later time or event. Use each of the sentences of (30) in a larger sentence which makes that connection clear.

2. Use the frame *the boy study Spanish* and give the sentence in the present perfect tense, the past perfect tense, and the future perfect tense.

3. Change the subject to *the boys* and do the same thing. What changes do you have to make?

4. The name of the present perfect tense may seem confusing to you, since it expresses a past event. Just keep in mind that it is called *present perfect* because it establishes a connection between a past event and the present. In what sense is there a connection to the present in each of the following?

   The wind has knocked over that tree.

   The secretary has left.

   A bird has built a nest in the rafters.

5. Name the tense in each of the following:

   He had prepared for this exam.

   The novelist has lost her motivation.

   By Monday, the ship will have reached the island.

   They have surrendered.

According to the formulas we have given for expressing verb tense in English, there should be nine different tenses: each of the three times standing alone as a tense (simple tenses), and each of the three times combined with one of the aspects (complex tenses). You may feel that these provide more than enough opportunity to describe the time of an action, but English gives us still three more complex tenses. The two aspects are not mutually exclusive and may occur together in the same sentence. When they do, the perfect aspect comes first, followed by the progressive aspect, each with its own auxiliary verb. The sentences of (31) are examples of how we use both aspects at once.

(31) I <u>have been working</u> for a long time.
      She <u>had been living</u> in New Mexico.

You will notice that in the first, the auxiliary verb *have* is followed by the past participle *been*, expressing the perfect aspect. But *been* also serves as the auxiliary verb for the present participle *waiting*, expressing the progressive aspect. Time only appears on the first auxiliary verb, and *been* does double duty as part of the perfect and part of the progressive. Since there are three times, we have three more complex tenses:

(32) She <u>has been waiting.</u> (present perfect progressive)
      She <u>had been waiting.</u> (past perfect progressive)
      She <u>will have been waiting.</u> (future perfect progressive)

All in all, English verbs may occur in twelve different tenses:

SimpleTenses	Complex Tenses
simple present	present progressive
simple past	past progressive
simple future	future progressive
	present perfect
	past perfect
	future perfect
	present perfect progressive
	past perfect progressive
	future perfect progressive

Now you know why they call it tense!

DISCUSSION EXERCISE 4.13

1. Use the frame *the boy study Spanish* and give the sentence in the present perfect progressive, the past perfect progressive, and the future perfect progressive.

2. Change the subject to *the boys* and do the same thing. What changes do you have to make?

3. Put each of the sentences you have just created in a larger sentence where the use of the tense makes sense.

4. Identify the tense of the verb in each of the following sentences:

By morning, Sam will have been waiting for thirty-six hours.

My in-laws have been organizing a party for us.

Until the war, London had been a thriving metropolis.

He has not been anticipating any trouble.

Before yesterday, I had not thought about the trip.

WHAT MAKES UP A VERB PHRASE?

When you look at a verb in a sentence, regardless of its tense, you will see that it may occur with modifiers. The verb and its modifiers make up a constituent known as a **verb phrase**. The verb is the head of the verb phrase and the permissible modifiers depend on the subcategory of main verb. The most important subcategories of main verbs are **intransitive, transitive**, and **linking**. We have encountered all of these in earlier discussions, but not by name.

Intransitive verbs are those that can stand all by themselves in their phrases. They may have modifiers, but they don't require them. The sentences in (33) all have intransitive verbs.

(33)   The children laughed.

My heart stopped.

The tree swayed.

The roof collapsed.

We *could* add modifiers to these verbs, additional words that told something more about the action, such as in the sentences of (34).

(34)   The children laughed at the clown.

My heart stopped when I saw them.

The tree swayed in the wind.

The roof collapsed under the weight of the snow.

But intransitive verbs do not *need* additional modifiers. They may stand all by themselves and constitute their own verb phrase. So we can say that *laughed* in the first sentence of (33) is an intransitive verb, it is the head of its verb phrase, and it makes up the entire verb phrase.

Transitive verbs, on the other hand, require a following noun phrase, as in the sentences of (35).

(35)   The pitcher <u>threw the ball</u>.

His father <u>bought a new suit</u>.

Sally <u>sold vegetables</u>.

We talked about these required noun phrases in the preceding chapter: they are the direct objects. So, transitive verbs are those verbs that require direct objects, and it is sometimes said that transitive verbs transfer their action onto the direct object. If we removed the direct object noun phrases from the sentences in (35), the sentences would be incomplete. The transitive verb and its direct object make up the verb phrases in the sentences of (35). Again, we could include other modifiers in the verb phrase, but the only two that are required are the verb and the following noun phrase.

DISCUSSION EXERCISE 4.14

1. What is the verb phrase in each of the following sentences? Is the verb transitive or intransitive? How do you know?

   The bell rings at 5:00 p.m. every day.

   Those children play all afternoon.

   Cats catch mice by instinct.

   The cynic snickered.

   His answer surprised us.

   Keith caught a cold last week.

2. There are many verbs in English that can be intransitive or transitive, depending on the sentence. So, we often cannot say whether a verb is transitive or intransitive unless we see how it is used in a sentence. For example, compare the pairs of sentences below.

The window broke.	The paper tore.
He broke the window.	He tore the paper.

   Are *broke* and *tore* transitive or intransitive? How can it be argued that they may be both?

3. There is some difference of opinion about whether to call verbs like *eat* and *read* in the sentences below transitive or intransitive.

   My family eats at 6:00 p.m.

   She reads at bedtime.

   The argument for calling them intransitive is that they can stand all alone in their verb phrases. What is the argument for calling them transitive?

The third important subcategory of main verb is a *linking verb*. Linking verbs do what their name suggests: they link a subject with a description of that subject. The one linking verb that we have encountered so far is *to be*. When this verb is the main verb, it is linking, as in the sentences below:

(36)   The cow was contented.

   The beans are in the pot.

I am unhappy.

She has been the president.

There are a variety of constituents that can follow a linking verb, including adjectives, prepositional phrases, and noun phrases. Linking verbs never stand alone in their verb phrase since, after all, their function is to "link" the subject to something else. There are other linking verbs in English as well, most of them used to describe senses or perceptions. The sentences of (37) illustrate some additional linking verbs.

(37)   Mary <u>feels</u> tired.

My goldfish <u>seems</u> lethargic.

The soup <u>tastes</u> funny.

This beer <u>smells</u> sour.

He <u>became</u> angry.

This paper <u>looks</u> messy.

Their music <u>sounds</u> terrible.

The situation <u>appeared</u> hopeless.

DISCUSSION EXERCISE 4.15

1. If you are thinking that the verbs in the above examples are not always linking verbs, you are right. Some of them may be transitive, and some may be intransitive. Consider each of the underlined verbs and see if you can create another sentence in which the verb is either transitive or intransitive, not linking.

2. Tell whether the underlined verb in each of the sentences below is intransitive, transitive, or linking.

That man *appeared* in my dream.

I <u>felt</u> so happy today.

Linda <u>lost</u> her contact lens.

He <u>feels</u> your pain.

Do you <u>taste</u> the pepper in this stew?

The lawyer <u>seemed</u> nervous.

I <u>question</u> your motives.

We <u>smelled</u> smoke.

The room <u>smelled</u> smoky.

WHAT ARE NONFINITE VERB PHRASES?

So far in our description of the verb phrase, we have been assuming that the verb of the phrase carries tense, and so we can call it a **finite verb phrase**. But it is also possible to have a verb phrase in which the verb is in one of its nonfinite participle forms and there is no tense expressed. These are called **nonfinite verb phrases** and often occur at the beginning of a sentence. As you will see in the sentences below, nonfinite verb phrases are like other verb phrases in terms of their modifiers. They may have direct objects, prepositional phrases, or any other modifier that appears in finite (tensed) verb phrases.

(38)   <u>Having abdicated his throne</u>, the king felt relieved.
<u>Smelling the poison</u>, the princess refused to eat her soup.
<u>Dismissed from class</u>, the children ran in all directions.
<u>Having eaten dinner</u>, we were able to relax.

What is interesting about these nonfinite verb phrases is that they do not have subjects of their own; they have to "borrow" the subject from the rest of the sentence. We know from the rest of the sentence that the king abdicated his throne, the princess smelled the poison, the children were dismissed from class, and we ate dinner. One common nonstandard usage involves a nonfinite verb phrase that cannot borrow the subject from the rest of the sentence. These are the famous **dangling participles**. Below are some examples:

(39)   Running for the bus, my book fell in the mud.
Having eaten dinner, the turkey carcass was put in the refrigerator.
Dismissed from class, the parents picked up their children.
Worried about opposition, the editorial was censored.

The meanings of these sentences are not hard to figure out, but technically the nonfinite verb phrases "dangle" because the book didn't run for the bus, the turkey carcass didn't eat dinner, the parents were presumably not the ones dismissed from class, and the editorial was not worried.

DISCUSSION EXERCISE 4.16

1. What is the finite verb phrase in each of the following sentences? What is the nonfinite verb phrase?

    Having received another rejection letter, I gave up writing poetry.

    Discovering the hidden treasure, Kelly jumped for joy.

    Lowered onto the floor, the bench looked smaller.

    Insulted by his cousin's remarks, Peter left the room.

2. Which of these nonfinite verb phrases are dangling participles? Can you rephrase the sentences so they don't dangle?

    Having written the best poem, the prize was given to Mary.

    Being a sloppy writer, Ian's notes were hard to read.

Realizing that he had insulted her, Jeff apologized to Carol.

Expecting the worst, my grades surprised my parents.

WHAT IS SUBJECT-VERB AGREEMENT?

Now that we have a better understanding of noun phrases and verb phrases, we can talk about an important cross-referencing rule of English called **subject-verb agreement**. You might remember that in Chapter 2 we introduced the idea of the clause. A clause is a combination of a noun phrase and a verb phrase. The noun phrase of the clause is called the subject, and the verb phrase is called the predicate. All sentences, then, are made up of subjects and predicates. (To refresh your memory, turn back to Chapter 3, Discussion Exercise 3.8) The rule of subject-verb agreement says that the verb of the predicate must agree with the subject noun phrase in person and number. We already know what **number** means: if the subject is a singular noun phrase, the verb must be marked as singular; if the subject noun phrase is plural, the verb must be plural as well. All of the following sentences are nonstandard because they violate number agreement.

(40)   The boy save his money.

My plants is dying.

It are on the table.

His answers sounds ridiculous.

Person is also relevant to the agreement rule. *Person* refers to the role of the noun phrase or pronoun in the conversation: first person is the speaker (or writer); second person is the person spoken to; and third person refers to anything spoken about. In more concrete terms, the pronouns *I* and *we* are first person, *you* is second person, and any noun phrase or pronoun that is being spoken about is third person. All subjects have both person and number. For example, *I* is first person singular, *we* is first person plural.

DISCUSSION EXERCISE 4.17

1. What is the person and number of the subject in each of the following sentences?

All people need compassion.

You should take a break.

This dog is a stray.

We agreed to disagree.

I won't report it this time.

The answers are in the book.

My sister arrived last night.

2. Which of these would you judge to be in violation of the subject-verb agreement rule?

She bring her lunch every day.

They expects to leave tomorrow.

She dislikes dishonest people.

The soldiers is returning from war.

I does not agree with you.

We have a problem.

The rule of subject-verb agreement is not something to which we need to pay attention all the time, because in many cases the verb stays the same for all subjects, regardless of person and number. In the simple past tense, for example, we don't have to think about which verb goes with which subject, because the verb doesn't change its form. When do we need to pay attention to the rule? One case that does require special attention is the verb *to be*, because it changes more than other verbs do. In the simple present tense, for example, there are three different forms:

	Singular	**Plural**
First Person	am	are
Second Person	are	are
Third Person	is	are

To be also requires special notice in the simple past tense. Unlike any other verb in the language, it has two different forms from which we must choose.

	Singular	**Plural**
First Person	was	were
Second Person	were	were
Third Person	was	were

DISCUSSION EXERCISE 4.18

1. What makes these violations of the standard English rule of subject-verb agreement?

They was laughing at me.

You is supposed to be my friend.

His parents is coming for dinner.

We wasn't invited to the party.

2. Give the standard English version of these sentences. Are there any differences in meaning between the standard and the nonstandard version of each sentence?

Other than the verb *to be*, the only other place where the rule of subject-verb agreement shows some effect is in the simple present tense. As we have already said, most of the time the verb in the simple present tense is merely the base form. But if the subject is third person singular, we must add the suffix *-s* to the base form of the verb. *The girl sees the boy* versus *The girls see the boy*. (Note that *-s* marks nouns as plural, but verbs as singular!) Not every dialect of English adds the *-s* to the verb in the present tense. It only carries information that is already expressed in the subject, and it upsets what is otherwise a neat pattern. But standard English still requires us to use it. We usually have no trouble recognizing when a subject is third person, but we sometimes have trouble recognizing whether the subject is singular or plural.

One situation in which the number of the subject might be difficult to determine is where the subject contains more than one noun phrase. If it is a compound noun phrase with *and*, we treat the subject as plural: *The man and the woman are working*. But the agreement requirement is different if there are two noun phrases joined by (*either*) *or*, or (*neither*) *nor*, as in (41), often called a **disjunction**.

(41)   Either my sister or my brother is picking me up.

Neither the man nor the woman works here.

You or your friends are expected to show up.

Here we might be inclined to think of these subject noun phrases as plural also, but standard English requires us to view them separately and imposes the following rule: *the noun phrase closest to the verb determines the agreement*. If you look again at the sentences of (41), you will see how that works. Sometimes following this rule gives rise to sentences that are awkward to the ear, like *Either he or I am to blame*, and we might avoid the problem by rephrasing the sentence: *Either he is to blame or I am*, for example.

The next set of sentences illustrates another situation in which it may not be clear to us what the number of the subject is.

(42)   One of the professors requires a term paper.

Each of the children receives a gift at Christmas.

Although the subjects are *one of the professors* and *each of the children*, it is only the head of the noun phrase that determines the agreement. That means the verb must agree with *one* and *each*, both of which are singular. This is an especially difficult rule to follow for two reasons. First, the meaning of the subject may be plural, as in *each of the children*; second, we are used to thinking of the noun phrase right before the verb as the subject. This rule of agreement goes counter to our intuitions about how English works and so it is often violated, especially in speaking. The same problem arises if the head of the noun phrase is a singular noun, as in the nonstandard sentences in (43).

(43)   The range of answers were interesting.
       The intelligence of the children amaze me.
       The use of cameras are prohibited.

A third situation that goes counter to our intuitions about subject-verb agreement is in sentences that begin with *there*. Consider the pairs of sentences in (44).

(44)   A book is on the table.
       There is a book on the table.

       Three books are on the table.
       There are three books on the table.

The meaning of each pair is the same; the only difference is that in the second sentence of each pair, the subject has been moved behind the verb and the expletive *there* has been put in its place. No matter whether it has been moved or not, the subject noun phrase determines the agreement on the verb. But there is a strong temptation to treat *there* as if it were the subject. This results in nonstandard but very common sentences like *There's three books on the table*, where *there's* is a contraction of *there is*.

Subject-verb agreement is not a difficult rule for us to follow as long as it doesn't collide with other things we know and expect about English: we expect the subject to appear before the verb; we expect the grammatical number and the meaning number to be the same; we expect that if there is more than one noun phrase in the subject the verb will be plural. These are all reasonable expectations, and they work most of the time for English. But there are instances in which the standard rule of subject-verb agreement requires that we set aside those expectations and "figure it out" instead. We are more successful at this in writing, since we have time to reflect, and less successful in speaking, because we speak and formulate our thoughts simultaneously.

DISCUSSION EXERCISE 4.18

1. Each of these sentences violate the standard English rule of subject-verb agreement. Tell what the violation is and what leads people to make the error. What is the standard version of each sentence?

   Neither my mother nor my father are going.

   Either Sue or I are supposed to respond.

   There's too many people in this room.

   Each of you have to take the exam.

   One of the dogs keep barking all night.

   Everyone in my family attend the reunions.

   There's three reasons for rejecting this offer.

   Neither you nor I are to blame.

2. The above instructions are written in nonstandard English. Did you notice? Let's hope so.

3. Parallel to sentences like *Each of the children receives a gift* are sentences like

   All of the children receive gifts.

   Both of the children receive gifts.

   Why don't these sentences cause subject-verb agreement problems for us?

4. There are many nouns, called *collective nouns,* that are grammatically singular but have plural meaning, like *group, committee*, and *team*. In American English, they usually require a singular verb: *the group is meeting, the team is playing*. In British English they may have plural verb agreement: *the team are playing*. Even in American English there are some collective nouns that can be thought of as singular or plural, such as *faculty: the faculty is responsible, the faculty are responsible*. Can you think of others that allow this fluctuation in number?

5. In his *New York Times Magazine* column "On Language" (February 19, 1995), William Safire objects to President Clinton's statement "There's the talkers and there's the doers." What do you think Safire's complaint is?

6. This headline appeared in a recent newspaper:

   The enlightening power of the arts are overblown.

   What rule of grammar does it violate?

You have now crossed an important threshold in the study of English grammar. Noun phrases and verb phrases are the basic building blocks of sentences. Once you understand what they are made of and how they behave, you have a grounding in how the language works generally. Much of what we will say in the remaining chapters of the book will build on and deepen this understanding. In the next chapter we will consider pronouns, which in most respects can be considered a special kind of noun phrase.

REFLECTIONS

1. It may seem strange to you that only third-person-singular subjects require a suffix on the verb. In fact, there was a time when all forms of the verb required suffixes, but most of them were lost over time. You can still see remnants of these earlier suffixes in most versions of the Bible and in earlier literary writing: *thou hast* (you have), for example. You will also notice that the -(*e*)*s* suffix used to compete with -(*e*)*th* as the third-person-singular verb ending. Look up Portia's famous speech in Shakespeare's *The Merchant of Venice* that begins, "The quality of mercy is not strained. . . ." How does this speech illustrate that these two suffixes competed during Shakespeare's time?

2. In her column entitled "Word Court" in *The Atlantic Monthly* (March 1996), Barbara Wallraff tells us about the rule of *will* versus *shall:* "The traditional distinction made in England is that in the first person *will* has to do with willpower—that is, it denotes intentionality—and *shall* with simple futurity, whereas the second and third persons reverse the pattern." She goes on to explain the difference in meaning between *I shall drown; no one will save me* and *I will drown; no one shall save me.* Can you figure it out?

3. Another problematic use of the verb *to be* occurs in *if-then* contrary-to-fact statements. You'll notice that the *if* part of the construction uses the past tense of the verb:

   If I <u>had</u> a million dollars, I would quit my job.

   If you <u>saved</u> your receipts, you could return unwanted merchandise.

   But there is something strange about the standard English requirement for *be*:

   If I <u>were</u> you, I would not do that.

   What is unexpected about this verb form? Why do you think many people use the nonstandard *if I was you . . .?* What is your judgment of *if the weather was better, we could have a picnic*? The historical explanation behind all this confusion is that the *if*-statement doesn't really use the past tense. It uses another verb form called the *subjunctive,* which still occurs occasionally in English. It just so happens that the subjunctive has the same form as the past tense except for *be*. For *be*, the past tense has two forms, *was* and *were*, depending on the subject, but the subjunctive has only the form *were* for all subjects.

4. The distinction in meaning between the present perfect tense and the simple past tense can be very subtle, and in some cases they are used interchangeably. There is probably no difference at all between *I ate the*

pie and *I have eaten the pie*. But notice that if you want to pinpoint the action at a specific time in the past, a difference emerges. Suppose you wanted to add *yesterday*. Could you add it to both? Which sounds odd? Why do you think the two tenses are not equally compatible with expressions like *yesterday, last week,* and so on?

5. There is another subcategory of main verb called *ditransitive verbs*. These require both a direct object and a prepositional phrase. One example is *to put*.

    I put the book on the table.

    *I put the book.

    *I put on the table.

    Can you think of any other ditransitive verbs?

6. The verb *to act* can be either intransitive or linking. Consider the sentences

    He acts odd.

    He acts oddly.

    Describe some behavior of his that would fit each of these descriptions.

7. It is possible for nonfinite verb phrases to have their own subjects. These are known as *absolutes* and occur only in very formal English. Which are the absolutes in the following?

    The king having abdicated his throne, the peasants rejoiced.

    The students having passed their exams, the partying began.

    You will notice that in these there is no requirement that the nonfinite verb phrase be linked to the subject of the finite verb phrase.

8. Dangling participles are actually part of a larger category of **dangling modifiers**. Modifiers at the beginning of a sentence are supposed to connect to the following subject; if they can't, they "dangle." What makes each of these a dangling modifier?

    *Unhappy with the election results, another vote was taken.

    *Located a block from home, it was easy for me to get to work.

    There was a time that sentences like the following were considered to have dangling modifiers:

    Due to the storm, the school was closed.

    In what sense can this modifier be considered dangling? What is your own judgment about this sentence? Listen for this construction in everyday usage. Do you know anyone who considers it to be nonstandard?

9. There is always more to say about dangling modifiers. When they are at the beginning of the sentence, modifiers must "borrow" the subject. But when they are at the end, they may borrow the subject *or* the noun phrase closest to them. So, some sentences may have two meanings:

   I saw the boy looking through my window.

   Who was looking through the window? There is nothing grammatically wrong with this sentence, but you need to be aware in your own usage that it can have two different interpretations.

10. You will find people struggling with agreement problems in sentences like the following:

    The number of students is growing.

    A number of students are coming.

    Although they look very much alike, they have different structures. In the first, *the number* is the head noun and is singular, so the verb is singular. In the second, *a number of* acts as a determiner, comparable to *several*, and the head of the noun phrase is *students*. Since *students* is plural, the verb is plural. Try to explain this to someone without using grammatical jargon.

11. Some nouns are singular in meaning but plural in form and require the verb to be plural. Some examples are *scissors*, *pants*, *trousers*. Why do you think these nouns have plural form?

12. The following is from a modern novel:

    He shone his own flashlight upward then, so that two beams shone against the mast now. (David Guterson, *Snow Falling on Cedars* (New York: Vintage Books, 1995), p. 452.

    Does this author's use of the past tense of *shine* match your own?

13. A character in another modern novel asks

    Is that because you understand I have to ask if you or your sister know where Hawkins is? (Alice Hoffman, *Practical Magic* (New York: Berkley Books, 1995), p. 275.

    What grammar rule does this question violate?

PRACTICE EXERCISES   (Answers on p. 259)   _____

1. Identify the nonfinite verb forms in the following sentences. Tell whether each is a present participle, a past participle, or an infinitive.

   1. I need to ask you a few questions.

   2. She was looking for her sister.

3. Eric hasn't considered all his options.

4. They will be expecting you at seven o'clock.

5. Having been informed of the results, Mary regretted her actions.

6. John is driving his father to the airport.

7. To be truthful, I don't like your new haircut.

8. Judy can't leave yet.

9. Sitting in my living room, I can see his parked car.

10. By next week, we will have been living here one year.

2. Which of the following -*ing* words are present participles and which are gerunds?

1. My children enjoy visiting their grandparents.

2. They are painting the bedrooms today.

3. Ted was taking a walk.

4. You are being a nuisance.

5. Seeing is believing.

6. She has been listening to the radio.

7. You can learn this by memorizing the tables.

8. Canvassing the neighborhood is a good way to get votes.

9. Canvassing the neighborhood, I discovered a lot of Republicans.

10. He wrote a book about breeding horses.

3. Which of the following are generally accepted as standard and which are generally considered to be nonstandard?

1. Vickie has wrote a letter to her congresswoman.

2. He has ran the marathon two years in a row.

3. Lay down and close your eyes.

4. The baby has drunk all her milk.

5. I have never swum in this lake.

6. They pleaded with her to stop.

7. Jerry has went home.

8. We have seen that exhibit.

9. She has lain in her bed all day.

10. He swang the bat.

4. Identify all the verbs in the following sentences. Which are main and which are helping? Of the helping verbs, which are auxiliaries and which are modals?

   1. You can leave your shoes at the door.

   2. We are all happy for you.

   3. His watch is running fast.

   4. This man has no enemies.

   5. I have been expecting you.

   6. She would wait by the telephone for hours.

   7. Cathy has had a cold all week.

   8. Ruth has been having second thoughts.

   9. We must stop all this bickering.

   10. Must you leave the door open?

5. Take the frame *Mary go home* and create a sentence in each of the following tenses:

   simple present: _____

   simple past: _____

   simple future: _____

   present progressive: _____

   past progressive: _____

future progressive: _____

present perfect: _____

past perfect: _____

future perfect: _____

present perfect progressive: _____

past perfect progressive: _____

future perfect progressive: _____

6. Name the tense of the verb in each of the following sentences:

    1. She bought her ticket.

    2. I will have been waiting for an hour.

    3. We meet up north every year.

    4. He has agreed to go.

    5. She had already left.

    6. They will accept the invitation.

    7. You were grinning.

    8. I have been expecting you.

    9. They will have eaten by then.

    10. We had been studying.

7. Tell whether the main verb in each sentence is transitive, intransitive, or linking.

    1. The plane flew over the mountain.

    2. The pilot flew the plane.

    3. The passengers seemed calm.

    4. A flight attendant appeared with beverages and peanuts.

    5. One child felt sick.

    6. I saw the approaching storm.

    7. Everyone was frightened.

      8. A little girl sobbed quietly.

      9. The plane landed safely.

     10. We all cheered.

8. Which of these sentences have dangling participles?

      1. Realizing that he would be late, Mike began to run.

      2. Having left her keys at home, Barbara couldn't get into her office.

      3. Neglecting to floss daily, Ben's teeth began to rot.

      4. Fortified with vitamins, I drink plenty of whole milk.

      5. Expecting the worst, we were pleasantly surprised.

9. Which of the following violate the rule of subject-verb agreement?

      1. There's too much work to do.

      2. Each of the paintings are worth a thousand dollars.

      3. Either Tom or Fran have the key.

      4. Neither my husband nor I needs the money.

      5. Jim and Mary frequently entertain guests.

      6. The range of exercises are overwhelming.

      7. Both of the children are sick.

      8. Either Lois or her children know the answer.

      9. Neither my aunt nor my uncle are at home.

     10. There's too many cooks spoiling this broth.

10. Make up a sentence that meets each description. It doesn't matter what tense of the verb you choose.

      1. *may* expresses permission: _____

      2. *be* is a linking verb: _____

      3. *feel* is a transitive verb: _____

      4. *could* is a conditional: _____

      5. *must* expresses probability: _____

6. a modal softens a command: _____

7. *look* is a linking verb: _____

8. *look* is an intransitive verb: _____

9. *taste* is a transitive verb: _____

10. *bend* is an intransitive verb: _____

5

PRONOUNS

WHAT ARE PRONOUNS?

Pronouns are words that, for the most part, take the place of or stand for noun phrases. Sentence (1) contains a noun phrase as well as a pronoun.

(1) <u>The elderly man</u> sat on the bench until <u>he</u> was asked to leave.

The elderly man is a noun phrase and *he* is a pronoun that takes its place, in the sense that it enables us to avoid repeating the noun phrase. You can see that pronouns make communication efficient. It would be awkward and burdensome to have to repeat whole noun phrases in conversation: *The elderly man sat on the bench until the elderly man was asked to leave.* In fact, since we expect people to use pronouns to avoid repetition, when the noun phrase is repeated we tend to assume that it refers to two different things.

The noun phrase that the pronoun stands for is known as the **antecedent** of the pronoun. In sentence (1), *the elderly man* is the antecedent of the pronoun *he*. Normally, of course, the pronoun follows its antecedent. We use a noun phrase and then a pronoun to avoid repetition of the noun phrase. But there are some situations in which the pronoun can come first. In the examples below, the antecedent follows the pronoun (in at least one interpretation of the sentence):

(2) After <u>he</u> was asked to leave, <u>the elderly man</u> began to sob.
     When <u>she</u> saw the mess, <u>my mother</u> called the police.

There are some cases in which the antecedent of a pronoun is not a noun phrase, but rather a whole sentence, as in (3).

(3)   I hate to say <u>it</u>, but <u>you didn't pass the exam</u>.

Furthermore, there are some pronouns that do not have grammatical antecedents at all. Instead, they refer directly to the participants in a conversation. In the sentences of (4), *I*, *we*, and *you* are pronouns, but they do not
have grammatical antecedents. That is, they refer to the speakers and listeners and are not being used to avoid repetition of a noun phrase.

(4)   <u>I</u> question your sincerity.
      <u>We</u> enjoyed our vacation.
      <u>You</u> must knock before entering.

You will remember from Chapter 4 that these are called first- and second-
person pronouns; only third-person pronouns can have true antecedents.

DISCUSSION EXERCISE 5.1

1. Tell what the antecedent is for each of the underlined pronouns. Remember that
   some pronouns do not have antecedents.

   Ken bought a coat, but <u>it</u> doesn't fit.

   Although <u>he</u> doesn't know <u>it</u> yet, Jon is taking Kate to the prom.

   Ted and Lora said that <u>they</u> would come to the party.

   <u>I</u> am embarrassed to say <u>it</u>, but <u>you</u> have spinach in your teeth.

   <u>We</u> found two pennies and put <u>them</u> in the bank.

2. Some people say that pronouns take the place of *nouns*. Can you demonstrate
   that this is not correct?

3. Other lexical categories also have pro-forms—that is, words that are used to
   avoid repetition. They are not as varied and complex as the pronouns, so they
   don't lend themselves to similar in-depth study. Can you pick out the pro-forms in
   the following sentences? What repetition do they avoid?

   The fish is in the bowl and the snail is there also.

   When Kathy left the room, her brother did so too.

   The puppy is nervous, but being so doesn't affect his appetite.

4. You will remember from Chapter 3 that the word *it* also has a placeholder function when no other grammatical subject is available. In which of the sentences
   below is *it* a true pronoun and in which is *it* a placeholder? Could any be both?

   It has been snowing all winter.

   Put it in the closet.

   It is too hot to walk.

   It is easy to see that you're happy.

   It isn't the one I asked for.

PERSONAL PRONOUNS

You will see as we go along that there are many different kinds of pronouns, each with a different function. The ones that generally come to mind first are the **personal pronouns**. As we have already mentioned, these are distinguished by *person*; that is, they can be first person, referring to the speaker; second person, referring to the listener; or third person, referring to whatever is being talked about. They also have *number*: they may be singular or plural. For example, *I* is a first-person singular pronoun, *they* is a third-person plural pronoun. Some personal pronouns are also distinguished by gender. If I asked you to tell me what the third-person singular personal pronoun is, you would have to say *he, she,* or *it*. If I wanted you to be more specific, I would have to specify gender as well: pronouns that refer to females are feminine; those that refer to males are masculine; those that refer to things are neuter. The gender of English pronouns is called *natural gender* because the use of the pronoun corresponds to the sex of the antecedent. If you know a language other than English, you might be familiar with *grammatical gender*. In Spanish and French, for example, nouns are arbitrarily assigned gender; in English, *table* and *feather* are both neuter because they are inanimate things, but in Spanish and French they happen to have feminine gender (which affects the choice of adjective and article). English used to have grammatical gender as well, but it has been lost over time. When we use pronouns, we choose the ones that match their antecedents (or the conversational participants they stand for) in person, number, and gender.

The one other feature that distinguishes personal pronouns is **case**. *Case* refers to the function of the pronoun in a sentence, which is independent of the function of its antecedent. There are three cases in English: **subject** (or **subjective**), **object** (or **objective**), and **possessive**. Consider the underlined pronouns in the sentences of (5).

(5) He found the missing key.
     The idea intrigued him.
     His father was kind.

These pronouns are all third-person singular, but the first is a subject, the second an object, and the third a possessive pronoun.

DISCUSSION EXERCISE 5.2

1. Personal pronouns are distinguished by person, number, gender, and case. Fill in the grid below to show all the personal pronouns of English. Use the following frames to guide your choices:

_____ saw the boy. (subject)

The boy saw _____ . (object)

_____ cat is in the tree. (possessive)

	Subject	Object	Possessive
Singular			
First Person			
Second Person			
Third Person			
masculine			
feminine			
neuter			
Plural			
First Person			
Second Person			
Third Person			

2. You might remember that we encountered the possessive pronouns when we were talking about the determiners. *My, your, his, her, our,* and *their* are possessive pronouns that fall into the determiner system and modify the head noun of the noun phrase, in exactly the same way that possessive noun phrases do. Thus *his* in *his house* might stand for *that man's*, for example. But possessive noun phrases can also stand alone, and do not have to be determiners.

> This manuscript is <u>the essayist's</u>.

> The fault is <u>John's and Ted's</u>.

> This piece of property is <u>that woman's</u>.

If you want to replace these noun phrases with possessive pronouns, you must use the *long form* of the pronoun. Fill in the sentences below with the long form of the possessive pronoun that corresponds to the possessive pronoun determiner.

Example:	This is my book.	The book is <u>mine</u>.
	This is your book.	The book is _____ .
	This is his book.	The book is _____ .
	This is her book.	The book is _____ .
	This is its book.	The book is _____ .
	This is our book.	The book is _____ .
	This is their book.	The book is _____ .

3. Which possessive pronoun does not have a long form? Which short and long forms are the same?

If you look at the grid you have prepared, you will notice some interesting features of the pronoun system. One is that there is no difference between the second-person-singular and the second-person-plural pronouns. This is a relatively recent development in English, which used to have a separate set of second-person-singular pronouns—*thou* (subject), *thee* (object), and *thy* (possessive)—no longer used in modern English. If you observe people in conversation, you will see that we often still need this distinction even though modern standard English does not supply it. For many speakers of English, *you all* (or *y'all*) serves as the plural in casual speech (although for some it can be singular also). Some of us make do with *you guys* or *you people*, or even the sorely stigmatized *youse*, formed on the basis of analogy with noun plurals. Standard English is clearly deficient from the standpoint of second-person pronouns, so people devise strategies for bypassing the deficiency and communicating what they need to communicate. It is a principle of human language behavior that people do not like to sacrifice meaning to follow the rules of standard grammar and when standard grammar, fails us, we find other ways to get our meaning across.

Another characteristic of our pronoun system that often leads to awkwardness in usage is the absence of a gender-neutral third-person-singular human pronoun. The problem arises in sentences like those of (6):

(6)   If anyone wants to go, _____ should sign up now.
      You should see a doctor and ask _____ to x-ray your arm.
      Some person left _____ umbrella in the closet.

The pronoun in the blank has to refer to humans, so it cannot be the neuter pronoun *it*. That leaves us only the masculine and feminine pronouns, but in these sentences we don't know the sex of the antecedent. So what are our choices for filling in the blanks? This is a situation in which you find people resolving the dilemma in different ways. The eighteenth-century grammarians suggested using the masculine pronouns, arguing that they covered both sexes, in much the same way that words like *mankind* refer to all people. This was never a popular choice because to most speakers the masculine pronouns are simply masculine, but you will still see some grammatical purists adhering to this rule. In more recent years, the use of the masculine pronouns to include both sexes has been labeled sexist language and is frowned upon by the editors of many scholarly books and journals. Unfortunately, there is no fully acceptable substitute. Some people will use *he or she* (or *him or her* or *his or her* for object and possessive pronouns respectively). There are also some written variations of these like *he/she* or *s/he*. Anyone who tries to use these combinations discovers how awkward they can be, especially if the pronoun needs to be repeated several times. In scholarly writing, authors may alternate the use of masculine and feminine pronouns, sometimes with a prefatory explanation of their choice. Another option that is sometimes available to us, but not always, is to change the antecedent to a plural noun phrase. That solves the problem, because the plural third-person pronouns are gender-neutral, as we see in (7).

(7)   If people want to go, <u>they</u> should sign up.

When students work hard we should reward <u>them</u>.

Children must bring notes from <u>their</u> parents.

Probably the most popular choice among speakers of English for resolving this dilemma is to use the plural gender-neutral pronouns *they, them,* and *their,* even when the antecedent is singular. As you would guess, this usage is more common in speaking than in writing, but it seems to be gaining ground on all fronts in recent years and may soon be considered standard.

DISCUSSION EXERCISE 5.3

1. Why is the standard English absence of a distinction between a singular and a plural second-person pronoun a particular handicap to people who wait on tables in restaurants? How would you resolve the problem if you were a server?

2. How would you fill in the blanks in the following sentences? Explain your choices.

No competent lawyer would advise ____ client to lie on the stand.

When a child learns to speak ____ can make ____ wishes known.

Someone knocked, but I told ____ you weren't in.

People who travel abroad need to have ____ passports validated.

If you see anyone, ask ____ what ____ are doing here.

3. To solve the gender-neutral pronoun problem, some people have suggested that we introduce a new pronoun into English. One such suggestion is *em,* as in *if anyone calls tell em I'm asleep.* What do you think of this as a solution? Why do you think it hasn't caught on?

4. If someone told you that an anonymous donor gave all his money to the university, would you think that the donor might be female? Should you be able to get this meaning according to the eighteenth century rules for pronoun usage? Explain.

Probably the most serious usage problem in our pronoun system is the distinction between subject and object pronouns. You remember from Chapter 3 that we signal whether a noun phrase is a subject or an object by where we put it in the sentence. We do not have to change the form of the noun phrase. Therefore, *the girl* in (8) is a subject if we put it before the verb and an object if we put it after the verb.

(8)   The girl caught the ball.

Jeff saw the girl.

But our pronoun system says that not only do we need to locate the pronouns appropriately to mark their grammatical function, but we also must choose a different form of the pronoun for each function. If we change *the girl* in (8) to a pronoun, it will be *she* for the first sentence but *her* for the second: *she caught the ball, Jeff saw her.*

The standard English rule for subject and object pronouns is simple in principle: use subject pronouns for subjects and object pronouns for objects (direct, indirect, and objects of prepositions). Adult speakers of English apply this rule easily in simple frames such as the ones given in Discussion Exercise 5.2 to help you fill in the grid of pronouns. But let's look at some situations in which the choice seems to be harder. Suppose the pronoun is part of a compound noun phrase, as in (9).

> (9) He and I are building a fort.

He and *I* are both part of the subject, so standard English requires subject pronouns. But if you observe people's usage, you will hear other variations of this compound: *him and me, me and him, him and I.* That is, for compound subjects there is a tendency for some people to use at least one object pronoun. This usage is considered uneducated and is often corrected by adults: "It's not *him and me*, it's *he and I*." In an attempt to correct this, many people overcorrect, or hypercorrect, and say sentences like (10):

> (10) She is proud of he and I.

(10) is nonstandard, of course, because *of* is a preposition, and the pronoun following it is the object of the preposition.

DISCUSSION EXERCISE 5.4

1. To refresh your memory, make a list of all the subject and object personal pronouns, labeling them according to person, number, and gender. Which subject and objects have the same form?

2. Fill in the blanks with subject or object pronouns, according to the standard English rules of pronoun selection.

   You and _____ (I, me) are sitting together.

   ___(He, Him) and ___(she, her) will arrive soon.

   Deliver this letter to ___(him, he) and his sister.

   This is a story about ___(her, she) and ___(we, us).

   They spotted you and ___(I, me).

3. Would you say *this is a secret between you and I* or *this is a secret between you and me*? Explain your choice.

4. Even if someone says *Him and me are friends*, using object instead of subject pronouns, we still understand the pronouns to be subjects. Why do you think that is so?

5. Rather than signaling function, it may be that subject and object pronouns in compounds have come to signal levels of formality: object pronouns are casual, informal, and intimate, while subject pronouns are formal and maybe even stuffy. Does this fit your own perceptions of their usage? What is your image of the person who says *her and me took a trip* as compared to the one who says *she and I took a trip*? How do you judge the person who says *I saw he and she at the concert*?

Besides compounds, there are two other situations in which the standard English rules for pronoun usage are not universally followed, even by the most educated. One deals with the use of pronouns after the verb *to be*. Which of the two answers in (11) is standard, for example?

(11)   Which one is the thief? This is him.
                             This is he.

You will probably hear both, as did the eighteenth-century grammarians. Needing to find a rule that decided which was correct, the grammarians used Latin as a model; constructions such as these in Latin require the nominative case or, in terms of English grammar, subject pronouns. So, *this is he* is the strictly formal standard English answer to the question. Many of us follow this rule in formal situations, such as when answering the telephone (*this is she, this is he*), but we tend not to extend it to our everyday usage. How many of us in answer to the question *Where are the books I lent you?* say *Those are they on the table*? We are apparently comfortable with placing subject pronouns at the beginning of sentences, where we expect to find subjects, but our comfort level drops precipitously when we are asked to put subject pronouns somewhere else in the sentence.

The other situation in which the formal standard rule collides with our intuitive sense about how English works is in comparisons. Which of the two sentences of (12) is "correct"?

(12)   I am taller than him.
         I am taller than he.

The eighteenth-century grammarians were faced with the same question. They reasoned that there was an implied continuation of the sentence: *I am taller than he is tall*. Therefore, use whatever pronoun fits into the implied continuation. According to this rule, the second of the two choices in (12) is "correct." It is easy to see why speakers of English tend to favor the first choice. Unless you happen to have been told the rule for continuing the sentence in your mind, you would not be likely to figure it out for yourself, since putting subject pronouns at the end of the sentence seems highly unnatural to us.

DISCUSSION EXERCISE 5.5

1. Choose one of the two pronouns given to complete each sentence. Explain your choice. Would the larger context make a difference in your choice?

   Who is it? It is ___ (I, me).

   Which are my keys? Those are ___ (they, them).

   Are you happier than ___ (she, her)?

   Which one is David? This is ___ (he, him).

   You know more than ___ (we, us) about this.

2. Suppose you were one of the eighteenth-century grammarians involved in discussions about comparisons and pronouns. Could you make a plausible argument that *I am taller than him* should be correct? You were asked to do the same thing in Chapter 1. Your new knowledge of pronouns and pronoun usage should enable you to make a more convincing argument now.

3. Imagine yourself knocking on someone's door. They say *Who is it?* Do you say *It is I* or *It's me?* What is the effect of each? Do you think you would choose to violate the standard English rule? You might be interested to know that standard French uses the equivalent of *it's me, c'est moi.*

4. How did this author resolve the gender-neutral pronoun problem in the previous question? Do you think the author of a grammar book should resolve the problem in this way? What would you have done?

REFLEXIVE PRONOUNS

Reflexive pronouns are those pronouns that end in *-self* or *-selves*. In many ways, they are like the personal pronouns, but they are used only under certain conditions. One of those conditions is illustrated in the sentences of (13).

   (13)   Keith likes him.

   Keith likes himself.

You will notice that when we use the personal pronoun *him*, we know that *Keith* and *him* are two different people. But when we use the reflexive pronoun, we know that *Keith* and *himself* refer to the same person. So, one occasion on which we use the reflexive pronouns is when the subject and an object in a sentence refer to the same entity.

Another purpose of the reflexive pronouns is to provide contrast, as in (14), for example,

   (14)   My whole family is Republican, but I myself am a Democrat.

Here the reflexives do not function as pronouns in a technical sense, since they are not replacing a noun phrase, but rather are repeating it for the sake of emphasis or contrast.

A third use of the reflexive pronouns is illustrated in (15):

(15)   The women built the garage (by) themselves.

As you can see, here the reflexive pronoun carries the meaning of "alone, without accompaniment," again not technically a pronoun function since it is not replacing a noun phrase.

DISCUSSION EXERCISE 5.6

1. What is the function of the reflexive pronoun in each of the following sentences?

   Todd cooked dinner himself.

   Irene saw herself as an activist.

   The nurse gave himself an injection.

   The island itself is calm, but the surrounding seas are dangerous.

   I myself prefer cats.

2. There is some apparent overlap in the use of personal and reflexive pronouns, as illustrated below:

   As for me, I will vote my conscience.

   As for myself, I will vote my conscience.

   Do you detect any difference in meaning between the two sentences?

3. Many people use reflexive rather than personal pronouns in compounds, such as

   I accept this award on behalf of my wife and myself.

   My family and myself like to vacation at the lake.

   What do you think prompts people to use reflexive pronouns in these situations? What is your opinion of the grammatical status of this usage?

You have undoubtedly noticed that reflexive pronouns come in a variety of forms and that they agree in person, number, and gender with their antecedents. Every native speaker knows this and would never say ungrammatical sentences like those of (16):

(16)   *The boys blamed itself.

        *The fire extinguished themselves.

The full set of reflexive pronouns is as follows:

	Singular	**Plural**
First Person	myself	ourselves
Second Person	yourself	yourselves
Third Person		
Masculine	himself	
Feminine	herself	{themselves
Neuter	itself	

These pronouns are a good example of how people in their everyday use of English strive to make sense of its grammatical system. The formation of the reflexive pronouns generally follows a pattern:

Reflexive Pronoun = Possessive Pronoun (Short Form) + *Self* (Singular)
+ *Selves* (Plural)

But two of the reflexive pronouns are glaring exceptions to this pattern: *himself* and *themselves*. These use the object pronouns, not the possessives. Some people will discard these in favor of the forms that fit the regular pattern:

(17) He did it hisself.

They considered theirselves lucky.

These are stigmatized forms, but you certainly can't fault their logic!

There are other deviations from the standard set of reflexive pronouns in people's linguistic behavior. *-Self* and *-selves* distinguish singular from plural, but this is redundant information, since the number is already expressed in the first part of the pronoun. It is not uncommon to hear sentences like those of (18),

(18) We saw it ourself.

Let them do it themself.

in which *-self* is used as the generalized reflexive suffix. Again, this usage leads to a simpler system with no loss of meaning. Although they are judged ungrammatical at this time in the history of English, all of the reflexive pronouns in (17) and (18) are good candidates for becoming acceptable in the future.

DISCUSSION EXERCISE 5.7

1. Keeping the above chart out of sight, choose the standard English reflexive pronoun for each sentence.

The horse injured ‗‗‗‗‗‗.

Jim prides ‗‗‗‗‗‗ on his honesty.

I caught _____ in a lie.

Those people should consider _____ lucky.

We don't blame _____ .

2. Explain the grammatical reasoning behind the creation of the pronoun *theirself* as in *They built it theirself*.

3. Consider the following sentence:

    If a person wants to succeed in life they should take care of themself first.

    Although *themself* does not exist as a pronoun in standard English, it makes particular sense to use it in sentences like this. Why?

RECIPROCAL PRONOUNS

Reciprocal pronouns are similar to reflexive pronouns in that they are used when the subject and object refer to the same entity. The only difference is in the way the action is distributed. The sentences of (19) show the difference:

(19)   The lawyers respect themselves. (reflexive)

    The lawyers respect each other. (reciprocal)

As you can see, with reciprocal pronouns, the subjects are always plural and the action is distributed from one entity to another. There are only two reciprocal pronouns: *each other* and *one another*. In modern English, they are used interchangeably, but historically there was a difference between them. *Each other* was used for two entities, and *one another* was used for more than two. Thus at one time the sentences of (20) had different meanings:

(20)   The houses are close to each other. (only two)

    The houses are close to one another. (more than two)

This distinction reflects an older grammatical sensitivity to the difference between two and more than two that is now fading from English. (See Reflections 1 at the end of this chapter.)

DEMONSTRATIVE PRONOUNS

Demonstrative pronouns are easy to talk about because we have already encountered demonstratives as determiners. We know that there are four of them and that they indicate location relative to the speaker as well as number: *this, these, that, those*. As determiners, they modify a head noun in a noun phrase: *this book, that car*. As pronouns, they take the place of a noun phrase, as in the examples of (21):

(21)   This is an interesting fact.
       Who told you that?
       Shall I order these?
       I need those for my presentation.

Like first- and second-person pronouns, they may occur without grammatical antecedents and can refer directly to something in the context of the conversation.

RELATIVE PRONOUNS

Relative pronouns are used in constructions called **relative clauses,** which we will explore in detail in another chapter. Basically, a *relative clause* is a sentence that has been incorporated into another sentence. When a noun phrase in the larger sentence is repeated in the relative clause, the one in the relative clause is changed to a relative pronoun. Schematically, relative pronouns appear in the following context:

[NP antecedent   [relative pronoun]]
Sentence   Relative Clause

For example, I might want to tell you *I know the woman,* but you will not know which woman I mean unless I tell you more. I might try to fix that by adding a sentence, as follows:

(22)   I know the woman. The woman won the prize.

As described above, English provides us with a way to incorporate the second sentence into the first and to eliminate the repetition of the noun phrase. So, instead of (22), two choppy sentences with an ill-defined connection between them, we can change the second to a relative clause and incorporate it into the first. When we do this, we replace the repeated noun phrase with a relative pronoun. The resulting sentence is (23).

(23)   I know the woman who won the prize.

The noun phrase that the relative clause describes is called the **head** of the relative clause. In (23), *the woman* is the head of the relative clause, and *who,* of course, is the relative pronoun. There are five different relative pronouns in English, as illustrated by the sentences of (24):

(24)   We admired the boy who (or that) caught the fish.
       She approves of the man whom (or that) you intend to marry.

The child <u>whose</u> puppy ran away is sad.

The house <u>which</u> (or <u>that</u>) is for sale needs a new roof.

How do we know which relative pronoun to choose? One thing you will notice from the examples is that you need to pay attention to whether the head of the relative clause is human or nonhuman. Setting aside the pronoun *that* for the moment, we see that if the head of the relative clause is human, the relative pronoun is *who, whom*, or *whose*. If it is nonhuman, it is *which* (but see Reflections 12 at the end of the chapter). That still leaves the question of how we choose among the human ones. In (25), we unravel the relative clauses of (24) to help you see what determines the choice of human relative pronoun.

(25)   We admired the boy. <u>The boy</u> caught the fish (who).

She approves of the man. You intend to marry <u>the man</u> (whom).

The child is sad. The <u>child's</u> puppy ran away (whose).

You can see now that human relative pronouns are marked for case, just like the personal pronouns, and you choose the pronoun according to its function in the relative clause. In the first it is a subject, so we use *who*; in the second it is an object, so we use *whom*; and in the third it is a possessive, so we use *whose*. We see also from the examples of (24) that the pronoun *that* can be used in place of all the relative pronouns except *whose*.

DISCUSSION EXERCISE 5.8

1. Identify the relative clause in each of the following sentences. Then identify the relative pronoun and tell which noun phrase is the head of the relative clause.

   I met the clerk who sold you that car.

   The dog which you trained has gone berserk.

   Any person that agrees with you is a fool.

   The dentist whom you recommended accepted my insurance.

   She is the teacher whose class you visited.

2. The choice between *who* and *whom* is a difficult one for speakers of English because all relative pronouns are placed at the beginning of the relative clause regardless of their function. Normally, we identify subjects and objects by word order, but we can't do that for relative clauses. Instead, we have to mentally unravel the clause to see where the repeated noun phrase was before the clause became incorporated. Figure out whether Standard English requires *who* or *whom* in the following sentences.

Example:	The child_____ you scolded is crying.
Unraveled:	The child is crying. You scolded <u>the child</u>.
Answer:	whom

The dancer _____ fell will be out of the show.

We respect the people _____ you have chosen.

The woman _____ sold me this blouse works in another department now.

I pity the person _____ they convicted.

Let me be the one _____ congratulates you first.

3. For many speakers of English (but not all), a test for whether the relative pronoun should be *who* or *whom* is to see if you can leave it out. *Whom* can be omitted with no change of meaning, but *who* cannot. Try this test with the sentences of Exercise 2 above. Do you need to change any of your answers?

4. As you are probably aware, *whom* is dropping out of English, at least in informal, conversational English. *Whom* now carries with it an air of formality that seems inappropriate to most for more casual contexts. Rephrase the following sentences to illustrate the various strategies people use for avoiding *whom*. Do any of these strategies mark speakers as uneducated?

This is the actor whom the talent agent discovered.

I recognize the woman whom you insulted.

Tell me the people whom you would like to invite.

INTERROGATIVE PRONOUNS

Interrogative pronouns are those question words that ask about the identity of a noun phrase. You will be happy to learn that they are very similar to the relative pronouns in both form and principles of usage. Like the relatives, they are sensitive to the distinction between human and nonhuman. The three human ones are *who?*, *whom?*, and *whose?* The nonhuman ones are *what?* and *which?* (used when a choice is implied). The sentences of (26) illustrate their use:

(26)   Who is making that noise? (human subject)

Whom did you invite? (human object)

Whose is this? (human possessive)

What is making that noise? (nonhuman subject)

What do you want? (nonhuman object)

Which is better? (nonhuman subject—choice implied)

Which do you want? (nonhuman object—choice implied)

Like the relative pronouns, they always occur at the beginning of the sentence, so you need to unravel the question to see whether what is being questioned is a subject or an object. As you might well expect, the choice between *who* and *whom* is equally problematic for questions, and many speakers use *who* for both without serious consequences.

DISCUSSION EXERCISE 5.9

1. Fill in the appropriate interrogative pronoun as specified

    ___ did you see? (human object)

    ___ did he steal? (human possessive)

    ___ is going on? (nonhuman subject)

    ___ told you that? (human subject)

    ___ do you want? (implied choice—object)

    ___ can I trust? (human object)

    ___ is better? (implied choice—subject)

    ___ is his problem? (nonhuman subject complement—figure it out!)

    ___ shall I say is here? (human subject)

    ___ costs more? (implied choice—subject)

2. Some people use *which*? to question human as well as nonhuman noun phrases. What would be an example of this usage?

3. Notice that some of these pronouns can also be determiners. Which ones? Give some examples.

4. Interrogative *pronouns* are used to seek the identity of a noun phrase. Other interrogative words are used to elicit other kinds of information. What information is being sought in each of the following?

    Where is she?

    Why can't you do it?

    When does the movie begin?

    How does this work?

UNIVERSAL AND INDEFINITE PRONOUNS

Universal pronouns are words that represent all-inclusive noun phrases: *each, all,* and combinations of *every* with *one, body,* and *thing.* Grammatically, all of them except *all* are singular, even though their meaning is plural. Notice the singular verbs in the sentences of (27):

    (27)   Each has its own place.

              Everyone in my family is rich.

              Everybody likes music.

              Everything looks good.

They may themselves act as the antecedents of other pronouns, but because they are plural in meaning they are often referred to by plural pronouns:

    (28)   Everyone in my family is rich. <u>They</u> own several homes apiece.

Indefinite pronouns, as their name suggests, generally refer to indefinite entities or quantities. Included in this group are *some* and *any*, and the various combinations of *some, any* plus *one, body, thing. None* is also an indefinite pronoun, but shortens to *no* if it combines with something else. Quantities such as *many, several, enough, few,* and *less* also fall under this category, as does the indefinite *one*, as in *one ought to listen to one's elders.*

DISCUSSION EXERCISE 5.10

1. Identify the universal and the indefinite pronouns in the following sentences.
   Everyone needs love.

   I don't need anything.

   Nothing pleases you.

   Some may appreciate this.

   Few arrived on time.

   Do you have enough?

   One ought to respect the law.

   Nobody heard the news.

   She dropped something into the river.

   All can sing but none can dance.

2. Which pronoun would you use when *someone* and *anyone* serve as antecedents? Give sentences that illustrate your choice.

3. Many of the universal and indefinite pronouns can also serve as determiners. Pick three and for each demonstrate that it can be a pronoun or a determiner.

4. The word *all* can be a universal pronoun, a determiner, or a predeterminer. Illustrate this with example sentences.

At this point, with the understanding you have gained of pronouns and how they work, you can probably identify all the major players and actions in the sentence. In the next two chapters, we explore in depth those lexical categories that give added richness and variety to the basic components of sentences: adjectives and adverbs, and particles and prepositions.

REFLECTIONS

1. English pronouns used to have three numbers: singular, *dual,* and plural. There was a separate set of pronouns that referred specifically to two things. We have lost dual number in pronouns, but there are still some ways in which English grammar remains sensitive to the distinction between two and more than two. For example, there may be a secret *between* two people, but if more than two people share it the secret

is *among* them. Can you think of any other differences based on the dual-plural distinction? You may have trouble thinking of them because this number distinction has already been lost in many people's usage.

2. In some people's speech, certain inanimate objects are assigned female gender, such as ships and cars. Why do you think these objects are assumed to be female?

3. The following appeared in a letter to a newspaper advice columnist: "We love to eat out and enjoy trying new restaurants. Nine times out of 10 our server . . . refers to us throughout the meal as 'You guys.' We find this annoying. Even if we get excellent service, we will tip less if the server calls us 'You guys.' There must be others out there who feel as we do, or do you think we are too stuffy?" How would you respond to the writer?

4. The following appeared on a letter of recommendation form from a large midwestern university: "May we have your candid judgment of this applicant's qualifications and promise for the successful completion of their chosen graduate program." How might a grammatical purist react to this? What does it tell us about the acceptability of plural pronouns with singular gender-neutral antecedents?

5. If you wonder about the acceptability of competing grammatical choices, the best way to find out is to put them to the test. Try out various options, say, for gender-neutral pronoun replacement in papers you write for other courses and see how your instructors react to them. It's risky, but it's all in the name of science!

6. What do you think would prompt a prominent TV talk-show host to say "I couldn't be happier for you and he," or a prominent politician to say "I have great respect for he and his family?" (We are interested in the grammar, not the meaning.)

7. Listen for the usage of subject and object pronouns in popular music. Do you find any deviations from the rules of standard English? What is the effect of this usage?

8. Another pronoun usage problem occurs in double constructions of a pronoun and a noun phrase, as in <u>We Americans love our automobiles</u>. Which of these do you think are standard?

   Us tall guys can't find pants to fit.

   We girls should give a party for her.

   Take us loyal followers with you.

   You are so kind to we weary travelers.

9. The eighteenth-century solution to the comparison question is not entirely without merit even if it doesn't fit the way people actually use English. There are circumstances in which the choice of pronoun, subject or object, could make a difference in the meaning of the sentence. What is the implied continuation of the sentence in each of the following?

   Mary likes Tom more than me.

   Mary likes Tom more than I.

10. There are some verbs that are inherently reflexive; that is, they can or must be followed by a reflexive pronoun which has no particular meaning or function. *Behave* is one such verb: *The children behaved themselves.* Can you think of others?

11. One oddity of English relative pronouns is that we do not have a non-human possessive pronoun parallel to *whose*. *Which* can be a nonhuman subject or object, but not a possessive. What do we do if we want to say the following as a relative clause construction?

    I found the book. The book's cover is torn.

    Survey students in your class to see how they would say this. The most common response is to rephrase it so that it is no longer a relative clause: *I found the book with the torn cover*, for example. Is there a way to preserve the relative clause?

12. There are various editorial preferences concerning the use of *that* and *which*. Consult some grammar handbooks or style manuals to see if you can detect a pattern of preference. If they phrase their concerns in terms of restrictive and nonrestrictive relative clauses, you will probably need to wait until the end of the course to understand these explanations (Chapter 12).

13. Many people think that the question *Whom shall I say is calling?* sounds highly educated and formal. Does it meet the requirements of standard English? Try to unravel it to see whether the interrogative pronoun should be *who* or *whom*.

14. There is some disagreement about the number of *none*. Is it singular or plural? Would you say *None of them are invited* or *None of them is invited*? Consult several traditional grammar handbooks and ask a few people what they think. Does a consensus emerge from your findings?

15. An airline company sent out a mailer to its customers defending the reliability of its planes. The same type of plane is used by other airlines and, says the mailer, "they, like we, have total confidence in this workhorse of the industry. . . ." In what way is this sentence a violation of the rules of standard English?

16. A character in Toni Morrison's novel *Song of Solomon* says, "Don't you city boys know how to handle yourself?" (New York: A Signet Book, New American Library, 1977), p. 284. What makes this nonstandard? Why is its meaning clear nevertheless?

PRACTICE EXERCISES   (Answers on p. 261) _____

1. What is the antecedent of each personal pronoun in the following sentences? If the pronoun has no grammatical antecedent, to what does it refer?

   1. Max lost his book but found it in the garden.

   2. Leslie said that we should be friends.

   3. The agency called Alice and offered her a job.

   4. Don't do the exercise if you find it too difficult.

   5. My dentist charges too much.

   6. The classes met for two weeks, so I signed up for them.

   7. Joe lied to Sally, so she divorced him.

   8. If you find the missing letter, send it to me.

   9. When he heard the news, my uncle hugged me.

   10. When you get to England, give us a call.

2. Give the personal pronoun that corresponds to the description.

   1. first- person-singular subject

   2. third-person-plural object

   3. second-person-singular possessive

   4. third-person-singular feminine subject

   5. third-person-singular masculine object

   6. first-person-plural object

   7. second-person-plural object

   8. third-person-singular neuter possessive

   9. first-person-plural possessive

10. third-person-plural subject

3. Replace the underlined possessive noun phrases with their corresponding long possessive pronouns:

1. My watch is running slow.

2. She found your slippers under the bed.

3. The students turned in their assignments.

4. His villa is situated on a cliff.

5. No one wanted to answer her question.

4. Tell whether the following sentences are standard or nonstandard according to the strictest standard English rules for pronoun usage.

1. Bob and I will meet you in an hour.

2. It is an honor for my husband and I to be here.

3. You are much smarter than him.

4. Seth and me went fishing last week.

5. This is her in the photograph.

6. We met him and her at a party.

7. They were more frightened than we.

8. Him and I tied for first place.

9. Let this remain between you and I.

10. My son likes to visit her and her sisters.

5. Give the reflexive pronoun that corresponds to each description.

1. first-person plural

2. third-person-singular feminine

3. third-person plural

4. second-person plural

5. first-person singular

6. What function is served by the reflexive pronoun in each sentence?

    1. I found myself all alone.

    2. She herself is very tolerant.

    3. They went to the movies by themselves.

    4. He himself prefers to memorize his speeches.

    5. Those doctors consider themselves experts in bone repair.

7. Replace the relative pronoun *that* in each of the following sentences with *who* or *whom*, following the traditional rules of standard English.

    1. This is the person that lost his notebook.

    2. We contacted the builder that you recommended.

    3. Here comes the woman that I have been expecting.

    4. I pity the child that has been crying all morning.

    5. I pity the child that the others teased all morning.

    6. Are these the students that will be taking the course?

    7. He is the person that helped me.

    8. Is Jan the person that you saw behind the curtain?

    9. Philip is the student that scored highest on the exam.

    10. I learned it from the teacher that retired last year.

8. Fill in the appropriate interrogative pronouns:

    1. —————— did you have for breakfast? (nonhuman object)

    2. —————— went to the museum? (human subject)

    3. —————— is better? (nonhuman subject—implied choice)

    4. —————— can you trust these days? (human object)

    5. —————— did the storm destroy? (human possessive)

9. Identify the pronoun in each sentence (ignore the personal pronouns) and tell whether it is reciprocal, demonstrative, relative, indefinite, or universal.

1. Everyone can learn this sport.

2. The children know each other.

3. I admire the drawing that you did.

4. This isn't easy for the losers.

5. Phyllis didn't see anyone in the office.

6. The soldiers protected one another.

7. One shouldn't expect too much.

8. Each shall have a chance.

9. The citizens will obey the rules that are approved by the majority.

10. Nobody is blameless.

10. Identify all the pronouns in the following sentences and give as much information as you can about their form and function.

    1. Can somebody help me?

    2. What can I do for you?

    3. Whose is this?

    4. Everyone came to my party.

    5. She bought the computer that they suggested.

6

ADJECTIVES
AND ADVERBS

WHAT ARE ADJECTIVES?

Adjectives are more easily identified by their function than by their form. Their main job is to *modify* nouns. This means they may be used to provide added description to a noun for purposes of embellishment or to help distinguish it from other nouns. All of the underlined words in the sentences of (1) are considered adjectives.

(1)   The <u>tall</u> man left but the <u>short</u> one stayed.
      A <u>certain</u> woman is needed for this <u>complicated</u> job.
      A <u>mere</u> child could not solve this <u>amazing</u> mystery.

As you can see, there is no particular marking that makes a word look like an adjective, so they are not easy to identify merely by how they look. There are, however, certain derivational suffixes that turn roots into adjectives, and when one of these appears, the lexical category is transparent. Some of these are *-ive (pensive, native), -able (portable, communicable), -ible (responsible, edible), -al (rational, political), -ful (thoughtful, careful)*, and *-ish (boyish, childish)*. There are also some derivational prefixes that can turn a positive adjective into a negative one: *un-(happy), dis-(satisfied), in-(competent), ir-(regular), il-(legible), im-(mature)*.
Many (but not all) adjectives also have inflectional markings for **comparative** and **superlative**. The comparative form is used to compare two different nouns, and the superlative form is used to compare more than two (note again the dual-plural distinction). The sentences of (2) illustrate the comparative and the superlative of the adjective *smart*.

(2) Katy is smart, but Jill is smarter.

Of the three women, June is (the) smartest.

You see that the comparative inflectional suffix is *-er* and the superlative suffix is *-est*. It has probably occurred to you that not all adjectives may add these suffixes. In some cases, we must use the words *more* and *most* to indicate the comparative and superlative, respectively, as in (3).

(3) Dan is more responsible than Jack.

This is the most beautiful landscape painting in the museum.

How do we decide whether to use the suffixes or the words *more* and *most*? There are some general guidelines, although they are subject to change over time and there is a lot of fluctuation in people's usage. In present day English, the guidelines are as follows:

One-syllable adjectives add the suffixes: *taller, slimmer, coldest*.

Adjectives of three or more syllables use *more* and *most*: *more responsible, most enviable*.

Adjectives of two syllables add the suffixes if they end in *-y* (*happier, loveliest*); otherwise they use *more* and *most* (*more honest, most handsome*).

DISCUSSION EXERCISE 6.1

1. List some additional derivational suffixes for adjectives with examples of words that use them. Think, for example, about the suffixes that indicate nationalities.

2. Can you think of any other prefixes that turn a positive adjective into a negative one? How do we know which prefix goes with which adjective? Which is the one that people are likely to use if they're not sure?

3. There are some exceptions to the rule of adding *-er* and *-est* to one-syllable adjective roots, most notably the adjectives *good* and *bad*. What are their comparative and superlative forms?

4. You will find the most fluctuation in the comparison of two-syllable adjectives. What are your judgments about each of the following pairs? Does everyone agree?

    He is handsomer than my brother.

    He is more handsome than my brother.

    This is my happiest experience yet.

    This is my most happy experience yet.

5. The comparative and superlative markers *more* and *most* have negative counterparts. What are they?

We have been assuming in our discussion so far that adjectives generally lend themselves to comparison, but this is true only of one subset of adjectives, called **gradable adjectives**. *Gradable adjectives* have the capacity for degrees in their meaning, so not only can they be compared, but they can also be modified by other words, such as *very, quite, somewhat,* and *exceedingly,* called **intensifiers**. There are other adjectives, called **nongradable adjectives**, that do not lend themselves to comparison or modification. They are absolute in their meaning and have no degrees. *Perfect* is an example of a nongradable adjective. Something is either perfect or it isn't, and there are no steps in between. Other examples of nongradable adjectives are *married, pregnant, square, silent,* and *indestructible*. In theory, the distinction between gradable and nongradable adjectives is clear, but in practice it is often blurred. People have a tendency in their use of English to move adjectives from the nongradable to the gradable category, which entails some shift in their meaning. A good example of this is the word *unique*. Historically, it meant "one-of-a-kind, " which would make it nongradable. But people frequently use it with comparisons and modifiers, with a shift in meaning to "unusual," as in the sentences of (4):

(4)   This is the most unique shop in the city.
        Her idea is more unique than yours.

There are grammatical purists who object to this shifting of category and see it as one more example of the decay of the English language, but for the most part it is a common element of language change and part of the normal development of the language.

DISCUSSION EXERCISE 6.2

1. Which of the following adjectives are gradable and which are nongradable? Are there any that are debatable? *reversible, speculative, quiet, special, intentional, supreme*

2. The adjectives *married* and *pregnant* are often used in conversational English as gradable adjectives. Give examples of this usage and tell what shift of meaning occurs when each is used as a gradable adjective.

3. Are comparative and superlative adjectives, like *taller* and *smartest*, gradable or nongradable? Explain your answer.

HOW DO ADJECTIVES MODIFY NOUNS?

As we have said, the function of adjectives is to modify nouns. This function is carried out in a number of different ways. When the adjective is part of the same noun phrase as the noun, as in *the happy children, the wild west,* we say that the adjective is an **attributive adjective**. When the adjective occurs in its attributive function, within the noun phrase, it directly precedes the noun. If you look again at the formula for noun phrases that we gave in Chapter 3, you can now fill in another detail: an optional adjective before the noun.

But adjectives may also modify a noun from outside the noun phrase, as in the examples of (5):

(5) The cost is <u>excessive</u>.

The children are <u>happy</u>.

We considered the issue <u>unimportant</u>.

In these sentences, the adjective does not share a noun phrase with the noun it modifies; rather, it is part of the predicate (see Chapter 4) and is called a **predicate adjective**. Predicate adjectives are commonly linked to a subject noun phrase via a linking verb: these adjectives are called **subject complements**, as illustrated in (6):

(6) Our house seems empty.

The horses appear nervous.

His father sounds worried.

This soup tastes funny.

You were impressive.

Predicate adjectives can also modify object noun phrases, as in the third sentence of (5), or the sentences of (7):

(7) I found the news <u>unsettling</u>.

She judged the report <u>inadequate</u>.

The doctor pronounced the patient <u>cured</u>.

The adjectives in these sentences are called **object complements**. Notice that although they are next to the nouns they modify, they are not part of the noun phrase. To summarize, we may say that there are two adjective functions, attributive and predicate. Predicate adjectives are of two types, subject complements and object complements, depending on what they modify. (Note that subject and object complements can also be noun phrases, as was discussed in Chapter 3.)

DISCUSSION EXERCISE 6.3

1. Tell whether the underlined adjectives in the following sentences are attributive or predicate adjectives.

The <u>angry</u> mother scolded the child.

The music sounds <u>funny</u>.

I consider him <u>intelligent</u>.

He taught this <u>lazy</u> boy algebra.

You are <u>intimidating</u>.

We encountered an <u>enthusiastic</u> crowd.

The jury judged her <u>innocent</u>.

This stew tastes <u>terrible</u>.

My <u>elderly</u> aunt just died.

They have <u>talented</u> relatives.

2. For all those you labeled predicate adjectives, tell which are subject complements and which are object complements.

3. Most adjectives can be either attributive or predicate, but a few are restricted in their usage. What is the restriction on *sole, afraid, mere*? What is odd about the adjective *certain*?

4. The sentence *They are amusing children* has two different meanings. Under one interpretation, *amusing* is an attributive adjective. What is it under the second interpretation?

WHAT ARE ADJECTIVE PHRASES?

Like other lexical categories, adjectives can serve as the heads of phrases, accompanied by modifiers of their own. **Adjective phrases** function in exactly the same way that single adjectives do. The most typical modifier of the adjective is the intensifier, as in the sentences of (8):

(8) I am extremely upset.

This is somewhat unusual.

The news is quite shocking.

We are very excited.

In each of these sentences, the adjective phrase is a subject complement. It can also be an object complement, as in (9):

(9) He considered the questions very annoying.

or attributive, as in (10):

(10) That is a somewhat unusual request.

Adjective phrases can also be formed by completing an incomplete adjective. Some examples of these are illustrated in the sentences of (11):

(11) She was afraid <u>to respond</u>.

I am sorry <u>to upset you</u>.

They are adept <u>at lying</u>.

We are full <u>of hope</u>.

He is ready <u>for anything</u>.

The underlined portions of the sentences in (11) complete the meaning of the preceding adjective and are called **adjective complements**. Note that the term *complement* is unfortunately overused in traditional grammar, but it will help to remember that a *complement* completes something. Therefore, subject complements complete the subject, in the sense that they give you more information about it; similarly, object complements complete the object and adjective complements complete the adjective.

Another way that adjectives can form phrases is by "stacking up." Unlike other lexical categories, adjectives may appear in a string, as illustrated in (12):

(12) The big red house.

A thin blue line.

A tall dark handsome stranger.

DISCUSSION EXERCISE 6.4

1. What makes up the adjective phrase in each of the following sentences?

   I found a small glass bottle.

   These results are interestingly deceptive.

   She was tired of listening.

   Anne seems unusually cheerful today.

   A rather odd character appeared at my door.

   He is suspicious of doctors.

   You should be kind to others.

   The gardener planted a short round prickly bush in my courtyard.

2. What is the function of the adjective phrase in each of the above sentences: attributive, subject complement, or object complement?

WHAT ARE ADVERBS?

It is difficult to talk about **adverbs** as a category because several different things are given the name *adverb* in grammatical description. What most people think of when they think of adverbs is those words that modify verbs. This type of adverb tells where, when, or how an action is carried out. The underlined words in the sentences of (13) are adverbs:

(13) We saw the film <u>yesterday</u>.

Sharon flew <u>home</u> to see her folks.

She <u>graciously</u> accepted my invitation.

These are broadly categorized as adverbs of time, adverbs of place, and adverbs of manner, respectively.

Another type of adverb, one we have already encountered, is the *intensifier*. Intensifiers modify adjectives and form adjective phrases: *very quiet, quite sincere*. But intensifiers can also modify adverbs of manner, as in the following:

(14)   The dog ate quite greedily.

They responded very enthusiastically.

He typed amazingly quickly.

In each of these sentences, the verb is modified by an adverb that tells the way in which the action was performed. But each of these adverbs is itself modified by another kind of adverb, an intensifier.

A third common type of adverb modifies an entire sentence and is referred to as a **sentence adverb**. These normally occur at the beginning of a sentence and inject some commentary on the part of the speaker. A few examples appear in the following sentences:

(15)   Fortunately, it didn't rain the day of our picnic.

Consequently, Mary had to find another babysitter.

Moreover, I couldn't afford to spend the money.

Other common sentence adverbs are *therefore, however, furthermore*, and *nevertheless*.

Adverbs, then, perform four distinct functions:

modify verbs: time, place, manner
modify adjectives: intensifiers
modify other adverbs: intensifiers
modify sentences: sentence adverbs

DISCUSSION EXERCISE 6.5

1. Identify the adverbs in the following sentences and tell what kind they are. Are there any sentences that have more than one adverb?

Amazingly, we got all the answers right.

Betty searched laboriously through the manuscripts.

You are being very silly.

I'll see you tomorrow.

However, the value of your stocks are down.

The politician spoke quite sincerely.

Stay here for the night.

The child sobbed uncontrollably.

Truthfully, I don't want to help you.

Slowly, Maxine approached the bobcat.

2. Did you notice the subject-verb agreement error in the above exercise? Let's hope so.

3. Adverbs have somewhat more order flexibility than other lexical categories. What other positions could the underlined adverbs occupy without changing the meaning of the sentence or making it sound odd?

<u>Therefore</u>, Al can't attend the lecture.

Cary inched <u>carefully</u> around the debris.

Mildred listened <u>attentively</u> to his arguments.

4. What kind of adverb is *easily* in *Easily, she lifted the child from its crib*? Explain your choice.

5. In formal standard English, the words *real* and *right* are not intensifiers. Give some examples of (nonstandard) sentences that use these words as intensifiers.

Although the similarities are not overwhelming, many adverbs share certain common properties. For one, adverbs of all categories may have the derivational suffix -*ly*, which is added to an adjective root to form the adverb. This -*ly* is a remnant of the Old English suffix -*lic* ("like," pronounced "*leech*"). So *happily*, for example, is historically *happy-like*, as in *She laughed happy-like*. Another common property is that, like adjectives, many of them can be made comparative or superlative:

(16) She runs quickly, but he runs more quickly.

My aunt worked most diligently of all to keep the family together.

You will notice that adverb comparatives and superlatives are typically formed with the words *more* and *most* rather than the suffixes -*er* and *est*. But this is just a consequence of the fact that most adverbs have more than one syllable. There are a few one syllable adverbs, and these do add the suffixes. Consider the sentences of (17):

(17) Ted works hard, but Gloria works harder.

Clint sings loud, but Charlie sings louder.

Jay runs fast, Pat runs faster, and Chuck runs (the) fastest of all.

These one syllable adverbs (*hard, fast, loud, slow*) are known as **flat adverbs**, and have exactly the same form as their corresponding adjectives.

DISCUSSION EXERCISE 6.6

1. Give sentences that contain the specified forms: the comparative of *fast, merrily, readily*; the superlative of *loud, charmingly, noisily*.

2. For each function, give an example of a sentence with an adverb that ends in *-ly* and

     modifies a verb

     modifies an adjective

     modifies another adverb

     modifies a sentence

3. Is the underlined word in each sentence an adjective or an adverb? How do you know?

     The choir sings too <u>loud</u>.

     The band sounds too <u>loud</u>.

     The men work <u>hard</u>.

     The rock feels <u>hard</u>.

     The race car is <u>fast</u>.

     The athlete runs <u>fast</u>.

     The turtle seems <u>slow</u>.

     The turtle walks <u>slow</u>.

4. The flat adverbs are irregular in that they don't behave like other adverbs. Do you have any evidence that people are trying to make any of them regular? Has standard English accepted any of the regularizations?

IS ALL WELL AND GOOD?

One of the problem areas in the use of adverbs involves the adjectives *good* and *bad* and their corresponding adverb forms. As we all know, they do not follow the regular pattern. The standard English pattern is as follows:

		Comparative	Superlative
Adjective:	good	better	best
Adverb:	well	better	best
Adjective:	bad	worse	worst
Adverb:	badly	worse	worst

The sentences of (18) illustrate the pattern.

(18) This cake is good, this one is better, and this one is best of all.
        This car runs well, this one runs better, and this one runs best of all.

        This cake is bad, this one is worse, and this one is worst of all.
        This car runs badly, this one runs worse, and this one runs worst of all.

Perhaps because the forms and meanings are so similar, there is some tendency to confuse the adjective and adverb forms. The most stigmatized version of this is the use of the adjectives *good* and *bad* for their corresponding adverbs, as in (19).

    (19)   That child reads good.
             That child spells bad.

This confusion is aggravated by the fact that the word *well* can also be an adjective, meaning "in good health," and has the same comparative and superlative forms as *good*:

    (20)   I was ill yesterday, but I am well today.
             I was ill yesterday, but I am better today.
             I was ill Saturday, better Sunday, and best of all on Monday.

DISCUSSION EXERCISE 6.7

1. Tell whether the underlined word is an adjective or an adverb.
    The music sounds <u>good</u>.
    The weather is <u>better</u> today.
    Jack feels <u>better</u> today.
    Johnny reads <u>better</u> this year.
    The cheese tastes <u>bad</u>.
    He reacted <u>badly</u> to the news.
    You sing very <u>well</u>.
    My mother is <u>well</u> today.
2. Which of the following are standard English? Which aren't? Explain your answers.
    I feel so badly about the accident.
    The soup tasted well.
    The machine runs well.
    Mary never danced bad.
    He did good on his exam.
    As a volunteer, he does good for the community.
    Jan felt good about her party.
    Does the music sound good?

Many speakers of English are sensitive to the stigma associated with the use of *good* and *bad* for *well* and *badly*, but we have difficulty sorting out standard usage, probably because of the similarity of the structures in which they appear. As you probably noticed in the exercise above, there are two typical verb phrase patterns in English, illustrated as follows:

> **Action Verb + Adverb:**   run quickly, dance well, work hard
> **Linking Verb + Adjective:**   feel happy, be ready, seem good

You will recognize the second formula as the subject complement construction. We often rely on word order to help us decide the choice of word in English, but word order won't help us in this situation. Instead, we must pay attention to the type of verb that occurs in the verb phrase and choose the following word accordingly. Aside from not being able to rely on word order, we are met with a host of other obstacles in making this choice: *well* can be an adjective or an adverb; the flat adverbs and the adjectives look the same; some verbs can be action or linking, such as *feel*. As a linking verb, it is followed by an adjective: *feel good*. As an action verb, it is followed by an adverb: *feel the cloth gently*. It is not at all surprising that we find a great deal of variation in the use of adjectives and adverbs, especially with *good, bad, well*, and *badly*. As a result of all this variation, we will probably see some changes in standard English as time goes by. For example, many educated people say *I feel badly*. Technically speaking, this is nonstandard if *feel* is intended as a linking verb. But many people know that *bad* is nonstandard in *the child spells bad*, for example, and aren't willing to risk the negative judgment. This hypercorrection is so widespread that for many it is the only acceptable way of expressing this idea.

DISCUSSION EXERCISE 6.8

1. Explain how the following could be standard or nonstandard, depending on the intended meaning.

    He feels badly today.

    She looked excitedly.

    Tom sounds well.

2. What is your opinion of the statement: *I feel strongly about that*? Is it standard? Is there another way to say it? How do you think it is viewed generally?

3. Some speakers of English use *poorly* as the adverb corresponding to *bad*. Give an example of this usage. Do you ever hear it used as an adjective?

4. What is the difference in meaning between *he acts odd* and *he acts oddly*? What is the grammatical structure of each sentence?

5. Is the word *lovely* an adjective or an adverb? How do you know? What about *kindly* and *friendly*?

WHAT ARE ADVERB PHRASES?

An adverb may act as the head of a larger construction called an **adverb phrase**. Adverb phrases perform exactly the same functions as single adverbs but contain modifiers of the adverb. One type of adverb phrase that we have mentioned before contains an intensifier, essentially an adverb modifying an adverb, as in (21):

(21)   Lou speaks rather fast.

   Proceed very cautiously.

Adverb phrases may also be formed with **adverb complements**, words that complete the meaning of the adverb, as in (22):

(22)   Fortunately <u>for us</u>, the package arrived early.

   Mike works harder <u>than a beaver</u>.

   She walked (as) haughtily <u>as a queen</u>.

There are other constructions that perform adverbial functions that are neither adverbs nor adverb phrases. We will see some examples of these in the next chapter on prepositions and prepositional phrases.

DISCUSSION EXERCISE 6.9

1. Identify the adverb phrase in each of the following:

   Her cat rather quickly learned to catch mice.

   The deer runs faster than the antelope.

   This tailor very meticulously removed the stitches.

   The patient stared as blankly as a zombie.

   Fortunately for him, the meter reader never returned.

2. For each phrase you identified above, tell which is the head adverb and which are the modifiers.

REFLECTIONS _____

1. Why do you think the tendency is so strong to use both *more* and *most* along with the comparative and superlative suffixes (in children's speech especially): *this is more better; he's the most handsomest man in the world*? They were not always considered nonstandard. Why do you think the eighteenth-century grammarians judged these constructions to be ungrammatical?

2. Before English settled on *-est* as the superlative suffix, it had a number of other competing superlative suffixes, including *-ost* , *-ist*, and *-m*. Some people combined *-ost* and *-m* to form another suffix, *-most* (which looked just like the free-standing superlative word). Can you think of any modern-day words that use *-most* as the superlative suffix?

3. Words that end in *-ing* can have multiple functions. What is the lexical category of *talking* in each of the following?

    This is a talking mailbox.

    Talking relieves tension.

    I am talking about you.

4. There are interesting questions about the order in which adjectives are allowed to stack up. For example, it is more natural to say *the little red brick house* than it is to say *the brick red little house*. Can you detect any pattern in the preferred ordering?

5. In Old English, adverbs were formed by adding the suffix *-e* to adjectives. Eventually the *-e* dropped off and people replaced it with *-lic*, which later reduced to *-ly*. The flat adverbs are the ones that did not pick up *-lic* after *-e* dropped off. But users of English continue to bring these flat adverbs into the fold by creating new forms with *-ly*. So now *slow* and *slowly* are both adverbs, and *slow* will undoubtedly fall into disuse as an adverb over time. *Soft* used to be a flat adverb. Can you find any dictionary that still acknowledges its status as an adverb? What about *quick*? Are these flat adverbs more or less acceptable in their comparative forms: *softer, quicker, louder, slower*?

6. Do you think the stigma attached to using adjectives for adverbs has contributed to the creation of the new adverbs *slowly* and *loudly*? Explain. Ask a few educated people to judge the acceptability of *he runs slow* versus *he runs slowly*. Do you see any pattern in their responses?

7. Explain what this means: *We do well by doing good.* What are the grammatical categories of *well* and *good* in these sentences?

8. Some grammatical purists object to the use of *hopefully* as a sentence adverb, as in *Hopefully, more communication will lead to greater understanding*. Why do you think they disapprove of this usage? (Hint: compare it to *fortunately*.)

PRACTICE EXERCISES   (Answers on p. 264)   _____

1. Which of the following words are adjectives? For each gradable adjective, give the comparative and superlative forms: *salivate, spectacular, rigid, mossy, partially, incandescence, verifiable, bland.*

2. Make the following adjectives negative by adding negative prefixes: *literate, acceptable, fortunate, reversible, mobile, functional, connected, polite, distinct.*

3. Identify every adjective in the following sentences. Tell whether each is gradable or nongradable, attributive or predicate.

    1. The strong boxer toppled his weaker opponent.

    2. My only son became depressed after his favorite dog died.

    3. The foundation of this incredible structure is not stable.

    4. The angry cook tasted the ruined pudding.

    5. The pudding tasted sour.

    6. We sell frames that are circular, square, and rectangular.

    7. Her dear friend rented a little cabin in the woods.

    8. They considered the elderly man incompetent, but they were wrong.

    9. Your thoughtful gift made Jane happy.

    10. She seems incapable of change.

4. Identify the adjective phrases in the following sentences. What function does the whole phrase perform: attributive, subject complement, or object complement?

    1. They seem very nervous.

    2. The engine is about to explode.

    3. My sister considers my boyfriend extremely lazy.

4. We are pleased to inform you of the results.

5. The big colorful leafy tree was blown over by the ferocious wind.

6. She is inclined to tell the truth.

7. I consider your question highly impertinent.

8. He noticed a tiny little spot on the brand new rug.

9. Very generous people allow their overly needy friends to take advantage of them.

10. This old battered building needs renovation.

5. Identify the adverbs in the following sentences and tell what lexical category they modify. Sentences may have more than one adverb.

1. The president ruled quite responsibly.

2. Nevertheless, I can't float you a loan.

3. The scientists are extremely excited about the findings.

4. You arrived too late.

5. Yesterday I danced quite well.

6. It's not possible, therefore, to plan the trip.

7. You waited so patiently.

8. This day has been quite lovely.

9. The dog happily bit the bully.

10. The sofa will remain, however.

6. Choose the standard English word for the blank in each sentence. For which are both acceptable?

1. The engine sounds —————— (good, well).

2. The pony runs —————— (slow, slowly).

3. I feel ——————— (bad, badly).

4. You talk too ——————— (loud, loudly).

5. The class performed ——————— (well, good) today.

6. She looked ———————(good, well).

7. He did ——————— (good, well).

8. They worked so ——————— (hard, hardly).

9. The bread feels ——————— (soft, softly).

10. He doesn't feel ——————— (well, good).

7. **Pick out the adverb phrases in the following sentences.**

1. She quite impudently asked my weight.

2. Did they play loud enough?

3. I can guess your age very easily.

4. The cat approached rather tentatively.

5. Nan writes better than her brother.

6. The baby talked well for her age.

7. I answered as civilly as I could.

8. He sang surprisingly well.

9. They responded as quickly as possible.

10. She works extraordinarily fast.

7

PREPOSITIONS
AND PARTICLES

WHAT ARE PREPOSITIONS?

Prepositions are words, usually small words, that indicate the relationship of a noun phrase to the rest of the sentence. This may seem like a strange definition to you, given that we devoted a whole chapter to describing noun phrases and their functions without talking about prepositions at all. You will remember that noun phrases can be subjects, direct objects, indirect objects, and complements, and their position in the sentence tells us the particular roles they play in the sentence. What prepositions do is allow noun phrases to play many additional roles. For example, they may perform some adverbial functions, indicating time, place, or purpose of the action, as in the following sentences:

(1) The club met <u>during</u> the night.
    Let's have dinner <u>before</u> the show.
    She did it all <u>for</u> love.

The underlined words, of course, are prepositions. Prepositions can also help noun phrases perform certain adjectival functions, modifying other noun phrases. The sentences of (2) illustrate examples of this function:

(2) The house <u>on</u> the hill is haunted.
    The boy <u>with</u> Mary is her brother.
    That student is <u>in</u> my study group.

There is no pattern to how prepositions indicate noun phrase function, but native speakers of English have learned them by adulthood, so you do not have to memorize them. If you want to teach preposition use to speakers of other languages, then you have a problem. Consider how all the different prepositions in the sentences of (3) indicate location:

(3) We live <u>in</u> the United States.

We live <u>on</u> Beacon Street.

We live <u>at</u> the corner of Hollywood and Vine.

Consider also how the preposition *by*, as illustrated in the sentences of (4), can have several different functions:

(4) I'll be there by noon. (time)

Meet me by the fountain. (place)

He passed the exam by cheating. (means)

The house was built by my father. (agent)

DISCUSSION EXERCISE 7.1

1. Tell what noun phrase function is signaled by the preposition in each of the following sentences:

   We met them on the slopes.

   He felled the tree with an axe.

   The car disappeared over the hill.

   I'll see you at noon.

   She said to meet her near the fountain.

   It flies like an arrow.

   He went to the party with his brother.

   They prepared for their exams.

   The dog ran into the yard.

   Don't open a can of worms.

2. Preposition use is highly unpredictable and may even vary from region to region within the United States. Do you wait *on line* or *in line*? Do you become *sick to your stomach* or *sick at your stomach*? Are you aware of other fluctuations in preposition use?

WHAT ARE PREPOSITIONAL PHRASES?

Prepositions always occur with a following noun phrase (or pronoun), called the *object of the preposition* (see Chapter 3). Together they make up a constituent called a **prepositional phrase**. Grammatically, the prepositional phrase can perform two different functions in a sentence, adjectival and adverbial. In its adjectival function, it can be part of a noun phrase, modifying the head noun, as illustrated in (5).

(5)   The restaurant <u>on the corner</u> is my favorite.

It can also be a subject complement, as in (6).

(6)   My favorite restaurant is <u>on the corner</u>.

In its adverbial function, a prepositional phrase modifies a verb and often tells the time or place of the action, as in (7).

(7)   We met them <u>at noon</u>.
       We met them <u>on the corner</u>.

As you can see, the same prepositional phrase can perform different grammatical functions. It is also important to remember that the function of the preposition within the prepositional phrase is different from the function of the whole phrase. For example *on* in the above sentences consistently indicates that *the corner* is a location, but the function of the whole prepositional phrase changes from sentence to sentence.

DISCUSSION EXERCISE 7.2

1. Identify the prepositional phrase in each of the following sentences. Tell whether its function is adjectival or adverbial.

   The ashes blew into her eyes.

   Let's go to the movies.

   The people in this neighborhood are friendly.

   The ship is at sea.

   I don't believe the price of that sweater.

   Take me to your leader.

   Please arrive on time.

   A woman with a mysterious smile left this package.

   The accident was at this intersection.

   The purpose of this exercise is evident.

2. Explain why the following sentence can have two different meanings.

   The astronomer saw her colleague with the telescope.

Prepositional phrases, like other constituents, hold together in sentences: the noun phrase and the preposition together form answers to questions:

(8)   Where do you live?

       On the corner of Hollywood and Vine

> In Texas
>
> At the intersection of Main and Elm

Prepositional phrases also move around as a unit:

    (9) I saw him <u>near the fountain</u>.

         It was <u>near the fountain</u> that I saw him.

And they can often be replaced by single pro-forms such as *here, there,* and *then*.

But there are also grammatical forces that work to separate a preposition from its following noun phrase. Sometimes we need to move a noun phrase (or more likely a pronoun) to the beginning of its clause, specifically in questions and in relative clauses. Let's look again at those two cases. If I know, for example, that you could see someone, but I don't know the identity of that person, the question might form in my mind as:

    (10) You could see whom?

For ordinary questions, standard English requires that we move *whom* to the beginning of the sentence:

    (11) Whom could you see?

Now suppose the question in my mind is (12):

    (12) You were speaking to whom?

If I move only *whom*, I am left with (13):

    (13) Whom were you speaking to?

The preposition that is left behind is called a **deferred preposition**. Formal standard English frowns upon deferred prepositions; many of us are aware that there is a rule that tells us not to end a sentence with a preposition. The standard English solution is to bring the preposition forward along with the noun phrase, giving us (14):

    (14) To whom were you speaking?

We see a parallel situation with relative clauses, which also require the pronoun to be moved to the front of the clause. Suppose we wanted to combine the two sentences of (15) into one sentence with a relative clause:

    (15) I know the man. You were speaking to the man.

If we moved the relative pronoun alone, the result would be (16),

    (16) I know the man whom you were speaking to.

This leaves *to* behind as a deferred preposition. Again, formal standard English requires that the preposition be brought forward as well, giving (17):

(17)   I know the man to whom you were speaking.

DISCUSSION EXERCISE 7.3

1. Identify the prepositional phrases in the sentences of Discussion Exercise 7.1. Show that they are constituents by applying some of the constituent tests we described.

2. We have described the formal standard English rule for deferred prepositions, but we also know that deferred prepositions are not highly stigmatized and sound more natural in more casual uses of English. *Whom* is also highly formal, and using *who* in its place is acceptable in less formal English. That gives us a variety of ways to express the mental question expressed in (12). What are they?

3. For relative clauses, we have the additional options of replacing *whom* with *that* or leaving it out altogether. What are the various less formal options for expressing (15) with a relative clause?

WHAT ARE PARTICLES?

A **particle** is the second element of a two-part transitive verb. Particles are often confused with prepositions because the two word classes look very much alike. Many words can be either prepositions or particles, and they seem to occupy the same position in sentences. The sentences of (18) help to illustrate the differences between them:

(18)   Dorothy turned down the brick road.
       Dorothy turned down the invitation.

Down in the first sentence is a preposition. It forms a constituent with the noun phrase *the brick road* and so can stand as the answer to a question: *Where did Dorothy turn?* The whole prepositional phrase moves as a unit, giving us sentences like *It was down the brick road that Dorothy turned,* and it can be replaced with a pro-form as in *Dorothy turned there. Down* in the second sentence is not a preposition; it is a particle. You'll notice that *down the invitation* does not hold together as a constituent: it cannot be a freestanding answer to a question, nor can it move around together or be replaced by a pro-form. *Down*, rather, forms a grouping with *turned*: together they make up the two-part transitive verb. There are other single-word verbs that could take their place: *refused*, for example. If we were to capture the constituent structure of the two sentences of (18) visually, they would look like this:

Preposition: Dorothy turned [down the brick road].
Particle: Dorothy [turned down] the invitation.

Another interesting fact about particles is that they are permitted to move away from their verbs by a special rule called **particle movement**. Particle movement allows us to move the particle behind the object noun phrase, giving us another way to say the same thing. Thus, *Dorothy turned down the invitation* can also be said as (19):

(19) Dorothy turned the invitation down.

You notice that we do not have to worry here about the rule against deferred prepositions, because *down* is not a preposition in this sentence.

DISCUSSION EXERCISE 7.4

1. Tell whether the underlined word in each sentence is a particle or a preposition. How do you know?

   She looked <u>over</u> the contract.

   I found it <u>over</u> the rainbow.

   They ran <u>down</u> the street.

   He burned <u>down</u> the house.

   Turn <u>in</u> your assignment.

   Sit <u>in</u> your seat.

   She turned <u>off</u> the ignition.

   She turned <u>off</u> the beaten path.

   It flew <u>out</u> the window.

   I put <u>out</u> the garbage.

2. Demonstrate that the word *up* can be a preposition or a particle.

3. Explain in grammatical terms why the following sentence has more than one meaning:

   He slipped in the alcohol.

4. How does the rule of particle-movement work if the object noun phrase is a pronoun? That is, suppose instead of saying *She turned down the invitation*, we replaced *the invitation* with the pronoun *it*. Does it make a difference in the application of the particle-movement rule?

It may be disconcerting to find time after time that words can play more than one role in grammar. We are sometimes led to believe that grammar is a naming exercise, that all we need to do is learn what things are called and label them correctly. By now it should be apparent to you that that isn't how language works and that understanding the grammatical structure of a sentence involves analyzing how words relate to other words in the sentence. What something is called depends on the context in which it appears. As users of the language we have the extraordinary ability to sort out the various possible interpretations of words and sentences and choose the ones appropriate for the occasion on which they are uttered. As students of grammar, we need to be able to identify all the possible grammatical interpretations of words and sentences. We must also understand that the forces of change in our language can alter those analyses over time. In the next chapter, we explore in greater depth the fluid, dynamic nature of English that allows words to slip from one lexical category to another and allows its speakers to create new lexical categories.

REFLECTIONS

1. In Chapter 3 we said that indirect object noun phrases are preceded by the prepositions *to* or *for*: *I gave the book to Bill, I bought the book for Bill.* You will remember that some linguists have argued that these are merely examples of prepositional phrases and should not have a special designation as an indirect object. Is there any reason to single them out as different from other prepositional phrases?

2. Prepositional phrases can be embedded inside other prepositional phrases. How would you analyze the structure of the following sentence? Can it have more than one analysis?

   Put it in the drawer of the desk near the window.

3. The best way to learn how people judge deferred prepositions is to ask them. Ask several speakers of standard English to comment on the following sentences:

   For which doctor are you waiting?

   Which doctor are you waiting for?

   You might ask: Are they both grammatical? Would you use them in different settings? Does one seem more natural than the other?

4. Speculate about why the eighteenth-century grammarians ruled against deferred prepositions. What reasons can you come up with? Do these reasons collide with the fact that moved particles are acceptable?

5. The above question contains a deferred preposition. What is it? How else could the question be formulated?

6. Children may say things like *Take out it.* What have they not yet mastered about English?

7. There are some two-part verbs that do not use particles. The following sentence has two interpretations:

   We decided on the boat.

   Under one interpretation, *on the boat* is a prepositional phrase telling the location of the decision. Under the other *decided on* is a two-part verb with the meaning "chose." But there is no interpretation in which *on* is a particle. Can you demonstrate that it isn't a particle?

8. In addition to being particles and prepositions, some words, like *up, down, in,* and *out,* can also be adverbs. When they are adverbs, they can be replaced by the adverb pro-forms *here* and *there.* Can you give examples of these words used as adverbs?

PRACTICE EXERCISES   (Answers on p. 265)   _____

1. Which noun phrases (or pronouns) are objects of prepositions in the following sentences? (Note that the *to* that is part of the infinitive is not considered a preposition.)

   1. I like to take vacations in the springtime.

   2. My favorite times are at the seashore.

   3. This secret is between you and me.

   4. You are under no obligation to testify.

   5. I'll get there by hook or by crook.

   6. Like us, they eat dinner early.

   7. I can't wait until my birthday.

   8. He's waiting for the right moment.

   9. Let's meet after class.

   10. You can succeed by working hard.

2. What is the function of the prepositional phrase in each of the following sentences, adjectival or adverbial?

   1. The dog with the flea collar is scratching.

   2. My father takes a nap in the afternoon.

   3. Let's ask the guard in the red helmet.

   4. The pot is on the counter.

   5. The box under the tree is a present.

   6. You've opened a can of worms.

   7. She hammered the nails into the board.

8. By tomorrow, the job will be finished.

9. He stood before the judge.

10. My best suit is at the dry cleaners.

3. What are the two meanings of the following sentence? Which function of the prepositional phrase corresponds to which meaning?
   I found the child with the radar detector.

4. What are the prepositional phrases in the following sentence? Which are embedded within another prepositional phrase?
   The puppy crawled into a pipe under the house around the corner.

5. Give the most formal version of each of the following sentences by moving the deferred preposition and making other necessary adjustments.

   1. Who are you looking for?

   2. That's the book she was talking about.

   3. Is this the pot I'm supposed to cook it in?

   4. I need to call the friends I'm going with.

   5. Which actor are you standing in for?

   6. I fear the world we live in.

   7. Is this the channel that the news is broadcaast on?

   8. Which bench did they sit on?

   9. This is the hill we rolled down as children.

   10. Which cliff did she jump off?

6. Perform particle movement on each of the following sentences:

   1. She turned off the ignition.

   2. Linda took out the trash.

   3. The teacher brought in the books.

   4. The librarian looked up the address for me.

   5. Lois put on her new dress.

    6. Arsonists burned down the building.

    7. The children turned in their parents to the police.

    8. Please hand in your homework.

    9. They turned down our offer.

   10. Hand over your money!

7. Which of the underlined words are particles and which are prepositions? Could any be either?

    1. Sally brought <u>in</u> the mail.

    2. Ed looked <u>over</u> the newspaper.

    3. Put the flowers <u>in</u> the vase.

    4. The cashier rang <u>up</u> the wrong amount.

    5. Please turn <u>off</u> the light.

    6. Ethel put <u>on</u> her new hat.

    7. He looked <u>up</u> the street.

    8. They found <u>out</u> the truth.

    9. Let's jump <u>off</u> the bed.

   10. There's a pot of gold <u>over</u> the rainbow.

8. For which of the following has particle-movement applied obligatorily? Why?

    1. She put the cat out.

    2. The lawyer turned them in.

    3. Turn the engine off.

    4. Don't write him off.

    5. Please turn the volume down.

    6. He let himself in.

    7. I tuned her out.

    8. He put the shoes on.

9. They found it out too late.

10. Take them off.

9. Which of the underlined words are prepositions, which are particles, and which are adverbs?

1. Let's go <u>out</u> tonight.

2. Please take <u>out</u> the garbage.

3. He ran <u>out</u> the door.

4. Clyde hobbled <u>up</u> the street.

5. Look <u>up</u> and you'll see a helicopter.

6. Turn <u>up</u> the volume.

7. They called <u>in</u> a consultant.

8. We met <u>in</u> a bar.

9. Come <u>in</u> from the rain.

8

LANGUAGE USERS AT WORK: MULTIPLE MEANINGS AND NEW CONSTITUENTS

By now you are probably getting used to the idea that learning about grammar is a complex exercise that requires thinking about how people speak and understand their language. When we try to analyze a sentence or decide how to label words, we have to ask ourselves all kinds of questions about how the language works: Which words group together into constituents? Which constituents are nested within other constituents? What is the function of the whole constituent within the sentence? Which words are heads and which are modifiers? Which are cross-referenced with others in the sentence? Which can mean more than one thing? Which can belong to more than one lexical category or subcategory? More often than not, we find that there can be more than one answer to some of these questions. That is, native speakers of the language may understand what is said in more than one way. We have seen many examples of this in the preceding chapters. For example, the sentence *They decided on the boat* could be understood as *We chose the boat* or *We made a decision on the boat*. *The uncomfortable children's shoes* could mean that the children are uncomfortable or that the shoes are uncomfortable. We also know that the same word can be many different things: *run*, in *there's a run in the stocking*, is different from *run*, in *I run a mile every day, I run this machinery*, or *I've had a run of bad luck*. The difficulty we may have in analyzing sentences grammatically makes it all the more amazing that people do this kind of analysis routinely in the course of language use, sorting out the meanings and relationships of words instantaneously and discarding the possible analyses that are not relevant to the context. Language users, all of us, are incredibly talented grammarians.

Along with the ability to sort out the roles and relationships of words and constituents, we also have the ability to create new words and constituents, sometimes by moving a word from one category to another, sometimes by creating new configurations of constituents. We saw one example of this latter ability when we discussed relative clauses in Chapter 5. We may start with two separate sentences in mind, as in (1):

> (1)   The pet service walks my dog. I found the pet service in the *Yellow Pages.*

and reconfigure the constituents so that the second is incorporated into the first as a relative clause:

> (2)   The pet service that I found in the *Yellow Pages* walks my dog.

As we will see later in this chapter, we also have the ability to add together constituents of the same type to create a larger constituent of the same type. So the noun phrase *the woman* can be combined with the noun phrase *the man* to form a new noun phrase *the woman and the man.* Both of these processes are open-ended, in the sense that we can repeat them indefinitely, as we see in the stylistically unappealing but grammatically possible sentences of (3) and (4).

> (3)   This is the boy that owns the dog that chased the cat that ate the canary.
>
> (4)   A woman and a man and a boy and a girl and a cat and a dog climbed out of the Volkswagen.

This chapter is devoted to exploring in greater detail those creative, open-ended aspects of English that make grammatical analysis challenging and ordinary language use positively awe inspiring.

CROSSOVER FUNCTIONS OF WORDS: WHEN IS A NOUN NOT A NOUN?

At this point, we take it as a given that words may belong to more than one lexical category. Words may shift over time from one category to another through the process of conversion, or **category shift**, mentioned in Chapter 4, or they may accidentally end up being more than one part of speech as a result of sound change, like *hard* and *fast*. It has come up so often in our discussion of English that you will probably not be surprised at how widespread crossover is in English. Let's take a closer look at the multiple functions of words.

One very common crossover is from noun to verb. Speakers of English seem to have a particularly strong tendency to create verbs out of existing nouns, not always with the approval of grammatical purists. The effect is that many, many words in the language can be nouns or verbs: *run, set, hit, love, laugh* are just a few examples.

DISCUSSION EXERCISE 8.1

1. Use each of the above words in sentences to show that they can be nouns or verbs.
2. *E-mail* and *interface* have crossed over into the verb category. What is the evidence? Can you think of other cyberspace terminology that has undergone the shift from noun to verb?
3. Can these words be nouns *and* verbs? Why do you think there might be disagreement on this point? *Assist, transition, license, host, debut, chair, golf.*

We have already seen that many words, such as *up*, can be prepositions, particles, or adverbs, as in the following sentences:

    (5)   It ran <u>up</u> the wall. (preposition)
           He picked <u>up</u> the package. (particle)
           She looked <u>up</u>. (adverb)

Up can also be a verb, as in (6):

    (6)   He <u>upped</u> the ante.

and a noun, as in (7):

    (7)   She has her <u>ups</u> and downs.

DISCUSSION EXERCISE 8.2

1. What justifies placing *up* in all these different lexical categories: preposition, particle, adverb, verb, noun?
2. In how many different lexical categories can *down* be placed? What about *out*?
3. Show that *like* can be a verb, a preposition, or a noun. Can it be anything else?
4. Show that *well* can be a noun, a verb, an adjective, or an adverb. Can it be anything else?

There are many other crossover words, some of which we have already discussed. You will remember that the flat adverbs—*loud, slow, fast,* and *hard*—may also be adjectives. In this case, it was loss of endings that led to the same forms rather than some conscious crossing of one category to another. The effect for us now is the same, however. We cannot know what lexical category these words belong to unless we also know the context in which they are used. We have also noted that words that end in *-ing* may belong to at least three different lexical categories: adjectives, gerunds, and present participles. Words that end in *-ed* may be past tense, past participles, or adjectives. Nouns may often function as adjectives, as in *the* <u>brick</u> *house.* And, sometimes adjectives function as nouns, as in *the* <u>poor</u> and *the* <u>needy</u>*.* If we want to know which lexical category a word belongs to, we should not try to memorize it. Rather, we should look for clues and ask ourselves a number of key questions to figure it out: What is its function in the sentence? What inflectional endings may be attached to it? What is its relationship to the other words in the sentence?

DISCUSSION EXERCISE 8.3

1. Show how the word *dancing* can be an adjective, a gerund, or a present participle.
2. Show how the word *closed* can be a past tense, a past participle, or an adjective.
3. What makes *brick* an adjective in *the brick house*? What makes *poor* a noun in *the poor*?
4. Words can also shift from one subcategory to another. What are some examples of nongradable adjectives that have become gradable in common usage?
5. Sometimes proper nouns shift to the subcategory of common nouns, so we may refer to someone as a *casanova* or a *pollyanna*. Can you think of other examples?
6. Brand names of products are often susceptible to the shift described in 5: *Vaseline, Jello, Scotch Tape, Cellophane*, and *Linoleum* are all brand names that have assumed generic, common-noun meanings. Can you think of others?

AMBIGUITY: WHEN CAN A SENTENCE MEAN MORE THAN ONE THING?

Because words can have many different functions, we often find that a whole sentence can be interpreted in more than one way. We call this situation **linguistic ambiguity**. *Ambiguous* in ordinary language can sometimes mean "vague," but that is not how it is used in grammatical analysis. Here it means "having more than one interpretation." Consider the sentence in (8):

(8)   Visiting relatives can be fun.

There are two distinct meanings we can give to this sentence. On one reading, *visiting* is an adjective modifying *relatives* and the whole noun phrase *visiting relatives* is the subject of the sentence. On the other reading, *visiting* is a gerund, *relatives* is its object, and the whole gerundive phrase *visiting relatives* is the subject of the sentence. Although people who have not studied grammar would not describe the two different interpretations in this way, they will still recognize that this one sentence has two different meanings and might, if asked about its meaning, give **paraphrases**, as in (9):

> (9)   It is fun when relatives visit.
>
>   It is fun to visit relatives.

We are not always conscious of ambiguity in our normal use of language because we are so adept at choosing the interpretation that fits the context and discarding the others. Nevertheless, when adult native speakers are asked to reflect on their language, they can almost always uncover the ambiguities and will certainly see them once they are pointed out.

One of the benefits of studying English grammar is you can now explain why certain sentences are ambiguous. Ambiguity often results from one of three structural factors: modification, grouping, or grammatical relations. **Modification ambiguity** is illustrated in sentence (10):

> (10)   I stopped at the pleasant vicar's cottage.

Am I saying that the vicar was pleasant? Am I saying that the cottage was pleasant? *Pleasant* can modify either of those two nouns and so the sentence can mean either. **Grouping ambiguity** is illustrated by the now familiar (11):

> (11)   We decided on the boat.

We know from previous discussions that *decided on* can group together or *on the boat* can group together, giving two different meanings to the sentence. **Grammatical relations ambiguity** is illustrated in (12):

> (12)   The chicken is ready to eat.

Here, *the chicken* can be thought of as the subject of *eat*, or it can be thought of as the object of *eat*, two different grammatical relationships with the same verb.

DISCUSSION EXERCISE 8.4

1. Paraphrase each of the following sentences to illustrate its multiple meanings.

    She is a rare butterfly collector.

    He is too selfish to love.

    They slipped on the ponchos.

    It is too hot to eat (at least three meanings).

    Look at the messy child's handwriting.

    I saw him with my new bifocals.

They said he would leave Saturday.

Sensitive men and women subscribe to this magazine.

Boiling water can cause burns.

Find out how old Mr. Clark is.

Send me more beautiful pictures.

2. Explain what makes each of the above sentences ambiguous. Is it modification, grouping, or ambiguity of grammatical relations?

3. How would you feel if someone said of you "*I cannot praise her (him) too highly*"?

The kinds of ambiguity we have discussed above all result from more than one option in how the structure of the sentence is analyzed. For this reason, it is called **structural ambiguity**. Understanding structural ambiguity requires an understanding of how sentences are structured and how the various parts relate to one another. There is also a simpler kind of ambiguity that results from a word having more than one meaning—that is, two homonyms that happen to fit into the same slot in a sentence, as in (13):

(13)   She can't bear children.

Here, the word *bear* can mean "have" or it can mean "tolerate." There aren't alternative structures to the sentence, just different meanings for an individual word. This kind of ambiguity is called **lexical ambiguity**. Often in English a sentence may have both of these kinds of ambiguity, with several possible interpretations. Consider sentence (14), for example.

(14)   The lamb is too hot to eat.

As in sentence (12), *the lamb* can be the subject or the object of *eat*, but *hot* may also have two meanings: "elevated temperature," or "spicy." Therefore, there is at least one other meaning: someone can't eat the lamb because it is excessively spicy. We could also argue that *lamb* and *chicken* each mean two different things as well: a live animal or cooked food.

DISCUSSION EXERCISE 8.5

1. Explain the lexical ambiguity of each of the following sentences

Meet me at the bank.

I have to buy a pen.

The witches met for a spell.

Leave the sherry near the port.

The drill needs polish.

2. What makes the following sentences ambiguous? (Outrageous interpretations are encouraged!) Point out both the structural and the lexical ambiguities.

He turned over a new leaf.	Can they be fighting roosters?
The chicks peeped in the window.	This is a spelling bee.
She ran down her ex-boss.	That is a stupid pet trick.
They agreed on a match.	He left instructions for us to follow.
Do I have that right?	I visited a permanent hair removal clinic.
I hit the dog with the bat.	She loves twirling batons.

CREATING NEW CONSTITUENTS: HOW DO WE BUILD NEW STRUCTURE?

We have been looking at certain qualities of language use that require its users to exercise judgment about how a sentence is to be understood in a given context: words may belong to more than one lexical category, and sentences may mean more than one thing. These aspects of English reveal the analytic prowess of its users: In order to communicate effectively as speakers of English, we must always be engaged in some serious grammatical analysis, quickly weighing all the grammatical alternatives of what is said to us, choosing the ones relevant to the context, and discarding the ones that don't fit. If an alien lands and asks me to take him to my ruler, I probably will not fish around in my desk for a measuring stick. If someone who has invited me to dinner announces that the chicken is ready to eat, I will probably not think that their barnyard pet is waiting to be fed.

Still another dimension of language use is our ability to create new configurations of constituents. Sentence (1) reminded us of how we create relative clauses to indicate a modification relationship between two separate sentences. In this section, we will focus on another process that creates new constituents : **conjoining** (also called **coordination**).

Conjoining allows us to combine two constituents of the same type to create another constituent of the same category. The resulting constituent is called a **compound**. Almost any two like constituents can form compounds. Each of the sentences in (15) contains one:

(15)   The otters and the sea lions swam in the river.

He and I remained friends.

She washed and ironed her clothes before she left for vacation.

He ran a stop sign and hit a tree.

Their bones are healthy and strong.

She wrote the note slowly and deliberately.

I've seen him in and around the neighborhood.
It ran across the street and into a field.
When and how will you do that?
He brought the cake and I brought the ice cream.

In the first, for example, the subject of the sentence is the compound noun phrase *the otters and the sea lions*. The two noun phrases are *conjoined* noun phrases and form a *compound* noun phrase. Schematically it might look like this:

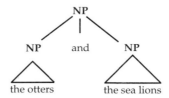

Another way to show how the two conjoined noun phrases nest within a larger one is like this:

[[the otters] and [the sea lions]   ]
         NP   NP   NP

The process is not limited to two, of course; any number of constituents can be conjoined, as illustrated in (16):

(16) The otters, the sea lions, and the walrus swam in the river.

He and I and our new spouses and their ex-spouses all remained friends.

DISCUSSION EXERCISE 8.6

1. What constituent types are conjoined in each of the remaining sentences of (15)?
2. Two lexical categories that do not lend themselves to *conjoining* (that is, cannot form compounds) are determiners and particles. Give examples of what these would be if English did permit it. Why do you think they don't form compounds?

When lexical categories are conjoined, they are held together by **coordinating conjunctions**. The most common one is *and*, which was used in all our examples so far. There are other coordinating conjunctions, all more restricted in their use than *and*. The underlined words in the following sentences are coordinating conjunctions as well:

(17) They are poor <u>but</u> happy.
       I am tired <u>yet</u> excited.

> Will you agree <u>or</u> disagree?
> She won't eat, <u>nor</u> will she talk.
> He was angry, <u>so</u> he left.
> The crowd dispersed, <u>for</u> the band never showed up.

You will notice that *but* and *yet* indicate something contrary to one's expectations. If you say *they are poor but happy*, or *poor yet happy*, you are implying that those two things don't normally go together. *Or*, sometimes called a **disjunction**, signals a choice among the conjoined elements: *I'll have pie or ice cream or frozen yogurt for dessert*. *Nor* is the equivalent of *and not*. There are also some two-part coordinating conjunctions, often called **correlatives**, that surround the first element of the compound, as illustrated in (18):

(18)   I need <u>either</u> a hammer <u>or</u> a saw.
<u>Neither</u> wind <u>nor</u> rain will deter the mail carriers.
<u>Both</u> Mary <u>and</u> Jim resigned from the committee.

Following are all the coordinating conjunctions of English:

Simple	Correlative
and	both . . . and
or	either . . . or
nor	neither . . . nor
but	
yet	
so	
for	

One particularly troublesome aspect of written English is differentiating between clauses (that is, noun phrase + verb phrase constituents) that are grammatically conjoined and those that are loosely connected to each other with a sentence adverb at the beginning of the second one. The effect is often not very different, but the standard English punctuation requirements are. Compare the sentences of (19) and (20):

(19)   The lawn is a mess, but I refuse to mow it one more time.

(20)   The lawn is a mess; however, I refuse to mow it one more time.
The lawn is a mess. However, I refuse to mow it one more time.

Coordinating conjunctions, like *and* and *but*, link the clauses together and belong to neither clause. Standard English punctuation permits a comma after the first clause, although omitting the comma is acceptable, especially if the clauses are short. On the other hand, sentence adverbs like those in

(20) belong to the second clause and are punctuated as part of it. (See Chapter 6 to review sentence adverbs.) The two clauses constitute separate sentences and are punctuated accordingly, with a semicolon or a period between them.

DISCUSSION EXERCISE 8.7

1. Give an example of each of the following:

    noun phrases conjoined by *or*

    verb phrases conjoined by *and*

    verbs conjoined by *or*

    adjectives conjoined by *but*

    adverbs conjoined by *but*

    clauses conjoined by *nor*

    nouns conjoined by *and*

2. What is the position of the coordinating conjunction if there are more than two items in the compound? Give examples.

3. What is the difference in subject-verb agreement between a compound subject joined by *and* and a compound subject conjoined by *or*? What is the rule? Use the sentences below to illustrate your answer.

    The dancer and the comedian ———— next.

    The dancer or the comedian ———— next.

4. Why might people find the following comments insulting?

    She is beautiful but bright.

    He is small but athletic.

5. Usually the correlatives do not signal any meaning difference from the corresponding simple coordinating conjunction, but *both . . . and* is sometimes different from the simple *and*. What is the difference in meaning between the following sentences?

    Alice and Tom are married.

    Both Alice and Tom are married.

6. Punctuate the following according to standard English punctuation rules:

    I studied therefore I did well on the test

    It was dry all summer so the crops didn't grow

    We attended the show but we missed the reception afterwards

    You need to turn on the machine furthermore you must boot up the disk

    Her remarks were unnecessary moreover I thought they were hurtful

    Our first stop will be Madrid and then we will go to Seville

    Which have coordinating conjunctions and which have sentence adverbs?

CLAUSE COORDINATION AND ELLIPSIS: ADDING AND SUBTRACTING FOR EFFICIENCY

The process of conjoining gives us enormous power to combine ideas when we communicate. But when we conjoin whole clauses we find a great deal of potential redundancy as well, as illustrated below:

(21)   The man left the bar and the woman left the bar.
       I washed the dishes and I dried the dishes.
       She wrote a poem and she wrote a short story.
       They bought the house and they renovated the house.
       Bob studied physics and Millie studied engineering.
       Cary bought the present and Steve wrapped the present.

When there is repetition in conjoined clauses, English allows us to eliminate the repeated elements without changing the meaning. This elimination of redundancy is called **ellipsis**. Each of the above sentences can be expressed in a shorter, more efficient manner:

(22)   The man and the woman left the bar.
       I washed and dried the dishes.
       She wrote a poem and a short story.
       They bought and renovated the house.
       Bob studied physics and Millie engineering.
       Cary bought and Steve wrapped the present.

No meaning has been lost, but certain material in each sentence has been omitted. (The omitted material may be referred to as *ellipted*.) When ellipsis occurs, the remaining elements may reorganize into compounds. *The man and the woman* in the first sentence of (22) is a conjoined noun phrase, for example. But you will notice that in the last two sentences of (22), no new compounds are formed. You will also notice that sometimes material is omitted from the first clause and sometimes from the second. As speakers of English, we always know which elements to leave out, but it would not be obvious to someone who was not a native speaker of the language. (See Reflections 13 at the end of the chapter for more on this.)

Clause ellipsis illustrates another dimension of the work people do when they are speaking and understanding English. We know that when we conjoin two clauses, we may also make them more compact by leaving out the repeated information. When we are listening to English, we also know which omitted information to restore so that we understand what is being said.

DISCUSSION EXERCISE 8.8

1. What has been omitted in each of the following? (In other words, restore the original conjoined clauses.)

   She studied mathematics and Al biology.

   Joe washed and Ed dried the dishes.

   Karen cut and sewed the dress.

   Martha Gray and her sister are married.

   I like pie and ice cream for dessert.

2. You will notice that the last sentence can mean two different things, but only one of the meanings can be derived from clause coordination followed by ellipsis. Which one?

3. Which of the following *cannot* be derived from clause coordination and ellipsis? Which have more than one interpretation?

   Ed and Nancy are happy together.

   Ed and Nancy are married.

   Ed and Nancy met in Reno.

   Ed and Nancy dance well.

4. What final adjustment must be made in conjoined clauses reduced by ellipsis of a verb phrase? Use the following as a test:

   The governor is studying the matter and the mayor is studying the matter.

In the remaining chapters, we will have more to say about clauses and sentences; in particular, the way they may combine with one another and the various functions they perform in communication. What we have emphasized in this chapter is that grammatical analysis is much more than taking pen to paper to label or diagram the pieces of sentences. Not only do we have to ask all the questions posed at the beginning of this chapter, but we also have to allow for multiple interpretations, and sometimes we must unravel compact and relatively efficient sentences to reveal their longer, more redundant counterparts in order to understand how people interpret them.

REFLECTIONS

1. In the recent past grammatical purists have objected to sentences like

   Contact his relatives.

   How will this impact your business?

   Loan me ten dollars.

   In each case, what do you think is the basis for the objection?

2. Why might people find the following sentence funny? [attributed to Groucho Marx]

   Time flies like an arrow, but fruit flies like a banana.

3. Check with several dictionaries to see if *golf* and *transition* are listed as verbs. Do you find agreement among them?

4. An author in a gardening magazine says that she "could wheelbarrow only 20 bricks at a time." A University of Michigan publication says, "Many other University organizations are partnering wth U-M Online. . . ." An Olympics TV announcer said, "No American has medaled in this event." An author of a scholarly article writes, ". . . most other participants will membership themselves as 'middle-men'. . . ." In what way is their use of English innovative?

5. Using nouns as adjectives is a highly productive process in English usage. Consider this: A holder for pencils may be called a *pencil holder*. Here, *pencil* takes on the function of an adjective. Labels for pencil holders might be called *pencil holder labels*. Here, *pencil holder* takes on the function of an adjective. What would you call a tray for storing pencil holder labels? Can you keep the process going? A real-life example of this process is the following: *Academic Plan Implementation Monitoring Committee.* Can you unravel its meaning?

6. If you tell your assistant to *xerox a contract and fax it to Smith Enterprises,* what kinds of category shifts are you participating in?

7. A proposed chain of motels was planning to call itself *McSleep Inns,* but was legally barred from using the prefix *Mc-* after the McDonald's Corporation claimed it was an infringement on their trademark. Are you aware of any generic use of the prefix *Mc-* ? What does it mean? Does it surprise you that a corporation can have trademark rights to a prefix? [If you want to know more about this case, read Genine Lentine and Roger W. Shuy, "Mc-: Meaning in the Marketplace," *American Speech* 65:4 (1990), 349–66.]

8. Do you think facial tissue companies other than Kleenex should be permitted to call their product *kleenex*? Why or why not?

9. A 1987 Supreme Court decision rested on the interpretation of the following instructions to a jury: "You must not be swayed by mere sentiment, conjecture, sympathy, passion, prejudice, public opinion or public feeling." The justices were divided on the question of whether *mere* modifies only *sentiment* or the whole list that follows. What do you think? If you want to know more about this case and its outcome, read

Lawrence M. Solan, *The Language of Judges* (Chicago: The University of Chicago Press, 1993).

10. Sometimes headlines exhibit ambiguity because they are so compact. *The Columbia Journalism Review* publishes examples sent in by readers. Here are just a few of the many that have been cited over the years. What is the ambiguity in each?

    Kicking Baby Considered to be Healthy

    Asbestos Suit Pressed

    Tuna Biting Off Washington Coast

    Criminally Insane Bill Passes

    Teachers Strike Annoying Students

    If you are interested in reading more about ambiguity, consult Steven Pinker, *The Language Instinct: How the Mind Creates Language* (New York: William Morrow, 1994).

11. What are the punctuation requirements for more than two conjoined items? How would you punctuate *She kept cats dogs lizards and hamsters in her apartment*? Consult several style manuals. Do they all agree on the punctuation rule?

12. Does standard English permit sentences to begin with a coordinating conjunction? Consult several style handbooks on this matter. Do you find different opinions? Did you notice whether the author of this book uses them at the beginnings of sentences?

13. There tends to be a pattern for clause ellipsis: elements on the right in their constituents are eliminated from the first clause; elements on the left in their constituents are eliminated from the second clause. Compare:

    Fred [caught the fish] and John [cleaned the fish].

    Fred [caught the fish] and John [caught the rubber tire].

    Look again at the sentences of (21) and (22). Draw brackets around the constituents containing the omitted elements. Does the pattern hold?

14. Sometimes transitive verbs may have implied objects, which gives the appearance of clause ellipsis: *I washed the dishes and Jim dried [the dishes]* These will not necessarily follow the pattern described in 13. Which of the following sentences follows the pattern for clause ellipsis and which merely has an implied direct object?

    Sue typed and Mary proofread the manuscript.

    Sue typed the manuscript and Mary proofread.

PRACTICE EXERCISES   (Answers on p. 267)   _____

1. What is the lexical category of each underlined word?

   1. Please <u>water</u> the flowers.
      Don't swim in the <u>water</u>.
      Are you fond of <u>water</u> sports?

   2. They're giving me a <u>baby</u> shower.
      I told you not to <u>baby</u> him.
      Hold the <u>baby</u>.

   3. We had to <u>drug</u> her.
      I work for a <u>drug</u> company.
      Take this <u>drug</u> for headaches.

2. What is the lexical category of *off* in each of the following?

   1. He jumped off the building.

   2. She ran off with her belongings.

   3. They chopped his head off.

   4. This is the off-switch.

   5. The gangsters offed him.

   6. I'm feeling a bit off today.

3. Tell whether the *-ing* word in each of the following is an adjective, a gerund, or a present participle. Could any be more than one?

   1. They are amusing children.

   2. Lying is wrong.

   3. You aren't listening to me.

   4. She is very intimidating.

   5. Entertaining friends is fun.

   6. Entertaining friends can make you happy.

   7. They are frightening us.

   8. We are buying some skiing apparatus.

    9. Screaming monkeys woke me up.

   10. I was arrested for loitering.

4. Name the lexical category of the underlined words:

    1. The news <u>upset</u> her.
She is very <u>upset</u>.
The news has <u>upset</u> her.

    2. You have <u>scared</u> the cat.
You <u>scared</u> her yesterday.
She is <u>scared</u> right now.

5. What kind of structural ambiguity is illustrated by each of the following: modification, grouping, or grammatical relations?

    1. She is a Russian literature teacher.

    2. We spotted the scientist with our telescope.

    3. I ran over the dog.

    4. Shooting hunters should be avoided.

    5. Rich accountants and lawyers belong to this club.

    6. They told us they were coming yesterday.

    7. He attended a little boys' school.

    8. Burning books sickened me.

    9. I heard that the news was good last week.

   10. He turned on a total stranger.

6. Consider the following two sentences:

Send me more beautiful pictures.
Send me less beautiful pictures.

The first is ambiguous, but the second isn't (at least not in formal standard English). Explain why not.

7. Name the compound category in each of the following:

    1. Rachel and Hannah are sisters.

2. Let's get together on or about the fifteenth.

3. They are rich but restless.

4. Please read this carefully and slowly.

5. I visited my grandmother and my grandfather.

6. Sue never studied yet she did well.

7. You and I will always respect each other.

8. Ed dropped off the package and left the room.

9. I cut and dried the flowers from my garden.

10. Look under the stove or next to the refrigerator.

8. Replace the simple coordinating conjunctions in the above sentences with correlatives wherever possible.

9. Tell whether the underlined word in each sentence is a coordinating conjunction or a sentence adverb. Punctuate the sentence accordingly.

1. The sky is cloudy <u>but</u> it won't rain

2. The class was delayed <u>for</u> no one had a book

3. My cousins were ill <u>nevertheless</u> they visited me

4. The roast was ready <u>so</u> I took it out of the oven

5. We had a good time <u>therefore</u> we exchanged phone numbers

6. Cats are affectionate <u>moreover</u> they are loyal

7. Our trip to Quebec was tiring <u>however</u> we enjoyed it

8. The elections were over <u>and</u> our party had won

9. I'm giving you the day off <u>furthermore</u> I'm increasing your salary

10. Turn yourself in <u>or</u> I will have to report you

10. Use the principle of ellipsis to eliminate redundancy in each of the following clause coordinations. Indicate which redundant material is omitted.

1. The boy fished in the pond and his sister fished in the pond.

2. Mary baked the potatoes and Mary baked the squash.

3. Bill bought a house and his brother bought a condo.

4. The squirrel ran up the tree and the squirrel ran down the tree.

5. All children need love and all children crave attention.

6. Her teeth are strong and her teeth are white.

7. The customer called the company and the customer threatened to sue.

8. All my friends came to our party and all her friends came to our party.

9. Convertibles are nice in summer and hardtops are nice in winter.

10. The man looked in the drawer and the woman looked in the cupboard.

11. What do the last two examples tell you about prepositions and ellipsis?

9

CLAUSE TYPE: VOICE

WHAT IS GRAMMATICAL VOICE?

There are two voices in English: the **active voice** and the **passive voice**. *Voice* has to do with the arrangement of the players, or actors, in the clause. The difference between active and passive voice is relevant when there are two players surrounding the action, a doer and a receiver. Normally, the order of elements in such clauses is doer + action + receiver. All of the sentences of (1) show this order.

(1)   An agent delivered the goods.
      My friend witnessed the accident.
      That pitcher threw the ball.
      The waiter poured the water.

You are already familiar with clauses of this type: you will recognize them as having transitive verbs with a subject and a direct object corresponding to the doer and the receiver. All the sentences in (1) are in the *active voice*. We have not had occasion until now to name this as the active voice, since this is the normal, expected line-up of grammatical relations and their meanings: subjects are doers and direct objects are receivers. (You might remember from our discussion of grammatical relations in Chapter 3 that subjects and objects actually play a wider range of roles than just doer and receiver, but for our purposes here, we can assume this narrower definition.) Most of our examples so far have been in the active voice and we have not had any reason to contrast them with other arrangements.

But there is another possible arrangement in English in which the order of elements is receiver + action + doer: this arrangement is known as the *passive voice.* We need to stop for a moment to consider what this means. Clearly, we don't mean that we are completely free to rearrange the noun phrases in a sentence. As speakers of English, we know that we cannot do that. Consider the sentences of (2), for example.

(2)   The dog bit the child.
      The child bit the dog.

If we merely have the doer and the receiver change places, we have a new sentence with a different meaning, and the subject is still the doer of the action. Both of these are in the active voice. If we want a clause to be in the passive voice, we have to signal that we intend for the first noun phrase to be understood as the receiver, not the doer, of the action. That is, we have to alert our listener (or reader) to intercept the normal interpretation and replace it with one in which the noun phrases don't have their normal or expected functions. We have several different ways in which we signal the passive voice, all evident in the sentences of (3), which are the passive versions of the sentences of (2):

(3)   The child was bitten by the dog.
      The dog was bitten by the child.

You will notice that there are certain systematic changes in structure as we go from the active voice to the passive voice.

DISCUSSION EXERCISE 9.1

1. Which of the following clauses are in the active voice and which are in the passive voice?

   The announcer reported the results of the election.

   A small child started the fire.

   He was frightened by the loud noise.

   The deer was killed by the hunter.

   My sister won the contest.

   The secretary shredded the documents.

   Basketball is enjoyed by everyone.

   My sister ordered this package.

   The game was stopped by the referee.

   The detective discovered the evidence.

2. Although we have not yet discussed the details of how we turn active into passive clauses, your native ability in English will enable you to do it instinctively. See

how quickly you can turn each of the following sentences into their passive-voice equivalents. (Make sure you preserve the meaning of the original active clause.)

The little dog buried the dirty old bone.

My nasty uncle Pete fired the servant.

The archeologists on the dig uncovered the ruins of the ancient city.

Sailors consume large quantities of fish.

Good music soothes our souls.

That teacher taught chemistry and physics.

The local newspaper published my letter complaining about trash collection.

The cat cornered the terrified mouse.

An honest stranger returned my lost wallet.

Your attitude surprises me.

HOW IS THE PASSIVE VOICE FORMED?

What is it that we are doing when we turn the active into the passive voice? One thing we do is switch the positions of the subject (doer) and the direct object (receiver). In fact, you will remember that one of the tests for direct objects described in Chapter 3 was the passive test, which involved placing the direct object in the position of the subject. But we have already seen that this exchange is not enough to signal the passive voice. What else do we do? We see that we also insert the preposition *by* in front of the doer of the action. This is how the elements line up in the passive as compared to the active voice:

Active Voice: Doer + Action + Receiver
Passive Voice: Receiver + Action + *by* **Doer**

In addition, we express the action in a different way. For the passive voice, we insert the verb *to be*, an auxiliary verb (see Chapter 4), and we put the main verb into its past participle form:

Active: **Doer + Action + Receiver**
                            |
                     tensed main verb

Passive: **Receiver + Action +** *by* **Doer**
                           to be **+ past participle of main verb**

DISCUSSION EXERCISE 9.2

1. Write out the passive versions of the sentences in Discussion Exercise 9.1. 2. What does it mean to say that *to be* functions as an auxiliary verb? What is its function?

2. To retain the same meaning from active to passive, the verb *to be* must carry the same tense as the main verb of the active sentence. Most of the examples we have given are in the simple past, but almost all tenses have corresponding passives. What are the corresponding passive sentences of the following?

   The proctor has administered the test.

   The whole town will remember you.

   The post office will have delivered the package by then.

   The researchers had expected those results.

   My cousin is building this house.

   The assistant was grading the exams.

3. What is the tense of *to be* in each of the passive sentences you have created above? (Refer to Chapter 4 if you need help on this.)

4. For most speakers of English, the following active sentences do not have corresponding passive forms. (Try to make them passive!) What do they all have in common?

   They have been forging many checks.

   The class will have been studying the complex tenses.

   The employer had been reviewing the files.

5. The verb *to get* may serve as the auxiliary instead of *to be* in the passive voice, usually in less formal contexts. Turn each of the sentences of Discussion Exercise 9.1.2. into its corresponding *get*-passive. Is there a meaning difference between the *be*-passive and the *get*-passive?

HOW ARE GRAMMATICAL RELATIONS DETERMINED IN THE PASSIVE VOICE?

One problem we face in talking about the passive voice is what grammatical terminology to assign to the noun phrases. In the active voice, it's easy. In a clause in the active voice, such as (4),

(4)   That company builds very fancy houses.

we can call *that company* the subject noun phrase because it is the noun phrase directly before the verb, it is the doer of the action, and the verb agrees with it in number and person. That is, it meets all our criteria for subjecthood. Similarly, *very fancy houses* is the direct object: it is the noun phrase directly following a transitive verb, it is the receiver of the action (in some loose sense), and it meets the passive test for direct objects.

But now let's look at sentence (5), the passive version of (4):

(5)   Very fancy houses are built by that company.

What shall we call *very fancy houses*? In some ways, it seems to be a subject: it is the noun phrase directly before the verb, and the verb agrees with it. But in another way, it is very unsubject-like: it is still the receiver of the action. Before we began to look at the passive voice, we could assume that a noun phrase is either a subject or it isn't. In the active voice the characteristics for subjecthood tend to converge on one noun phrase, just as we have described for *that company* in sentence (4). Similarly, the characteristics of direct object also tend to converge on one noun phrase, as we have described for *very fancy houses* in sentence (4). But in the passive voice, we see that these various characteristics get distributed in a different way: In sentence (5), *very fancy houses* is a subject by some criteria but not all; *that company* doesn't seem to be a subject in most respects, yet it is the doer of the action. If we are to describe the grammatical relations of the noun phrases in passive sentences, we have to fine tune our terminology to allow for this realignment of characteristics. We can do this by separating the purely formal, grammatical features of the noun phrases from their logical relationships to the action. In sentence (5), we say that *very fancy houses* is the **grammatical subject**, by virtue of its location before the verb and the fact that the verb agrees with it, but we say that it is the **logical direct object**, by virtue of the fact that it is the receiver of the action. What about *that company* in sentence (5)? We call it the **logical subject**, because it is the doer of the action, but since it forms a prepositional phrase with *by* we call it a **grammatical object of a preposition.** Naming noun phrase functions in the active voice is a straightforward matter of identifying the characteristics of each noun phrase and assigning it its function. The passive voice, on the other hand, requires that we recognize two kinds of functions, those that are purely formal (based on form or location), and those that are more meaning-based and tell the relationship of the noun phrase to the action. Consequently, noun phrases in passive clauses have two labels apiece. Schematically, the two voices look like this:

Active: Subject + Action + Direct Object
Passive: [Grammatical Subject + Action + *by* + [Grammatical Object of a Preposition
        Logical Direct Object   Logical Subject

DISCUSSION EXERCISE 9.3

1. We said that the verbs in (4) and (5) agree with their subjects. Demonstrate this by changing the subject in each so that the verb must also change.

2. What is the grammatical function of each noun phrase in the following active clauses?

   The reporter discovered two new pieces of evidence.

   A large number of people tolerate dishonest behavior.

Many potential employers reject sloppy resumes.

His students planted a memorial garden.

Thousands of tourists visited the museum exhibit.

2. Describe the functions of the noun phrases in the following passive clauses. Remember that each noun phrase will have two descriptions.

Two new pieces of evidence were discovered by the reporter.

Dishonest behavior is tolerated by a large number of people.

Sloppy resumes are rejected by many potential employers.

A memorial garden was planted by his students.

The museum exhibit was visited by thousands of tourists.

3. Normally, the two noun phrases that get reversed in the passive voice are the active subject and direct object. One exception to this is illustrated by the sentence below:

The clerk was given a second chance by the sympathetic customer.

What makes this sentence an exception to the reversal rule?

4. How does *the clerk* come to occupy subject position in the sentence above? What is its logical role? If you are having trouble answering this question, think of other ways that this sentence could be said. It will also help to review the section on grammatical relations in Chapter 3.

5. Name the function of each noun phrase in the sentences below. Some will have more than one description:

My father gave a dozen roses to my mother.

My father gave my mother a dozen roses.

A dozen roses were given to my mother by my father.

My mother was given a dozen roses by my father.

WHY DO WE NEED THE PASSIVE VOICE?

The passive voice adds nothing to the ability of English to express meaning. When we turn an active sentence into its passive counterpart, we rearrange the grammatical relations, but the meaning is left intact. This raises an interesting question: Why would a language have two ways of expressing the same meaning? Is the passive voice merely a redundant structure in the language, a frivolous and unnecessary addition that serves only to complicate the lives of students of grammar and contributes nothing to the expressive power of its users? Or, is there a legitimate function of the passive voice independent of the literal meaning it conveys? You are undoubtedly expecting a "yes" answer to this last question, so let's explore what a "yes" answer means.

Although there is no literal meaning difference between a clause in the active voice and its corresponding passive, there are still some differences, primarily of focus or emphasis. What the passive voice does for us is allow us to put the receiver of the action in a more prominent position in the clause and play down the importance of the doer. The subject position, the first noun phrase in the clause, is the most prominent. If we put the receiver there, it takes on more importance. Similarly, the doer of the action, by being made the object of a preposition at the end of the clause, is no longer prominent. Being able to realign the importance of subject and object noun phrases is a handy tool for us in communication. Suppose, for example, that we are having a lengthy conversation about your Aunt Tillie. If I ask you, "What finally happened to your Aunt Tillie?" You might answer,

(6) She was arrested by the police and sent to jail for mail fraud.

It wouldn't be wrong to say this in the active voice, as in (7):

(7) The police arrested her and sent her to jail for mail fraud.

But doing so would take the focus off Aunt Tillie and interrupt the continuity of the conversation. Stylistically, the passive voice is a useful alternative to the active voice, even though it doesn't change the meaning.

DISCUSSION EXERCISE 9.4

1. Construct a conversational context in which the following passive sentences would be an appropriate alternative to the active voice.

    The reports were found by the cleaning staff.

    My printer is being repaired by a computer expert.

    The bad news had been anticipated by most of us.

    Foreign automobiles are sold by that dealer.

    The textbook was chosen by the principal.

2. Students are sometimes discouraged by teachers from using the passive voice in their writing. Why do you think this is so?

3. Rephrase Exercise 9.4.2. in the active voice. Does this change how you would answer the question?

WHAT IS A TRUNCATED PASSIVE?

There is another very important difference between the active and the passive voice. Not only does the passive voice allow us to diminish the importance of the doer, it allows us to leave it out entirely. Consider the sentences of (8), for example:

(8)   The suspects were brought in for questioning.

Her car was wrecked in the accident.

His name has been eliminated from the list.

You are expected to arrive on time.

These are called **truncated passives** because the doer of the action is left unexpressed, in contrast to the **full passive**, in which the doer is expressed. In each of the clauses of (8), there is an implied doer, but we are not forced to express it in grammatical terms. The option of talking about an action and a receiver without expressing a doer is normally not available to us in the active voice, and that is what makes the passive voice truly useful to us.

Under what circumstances would we prefer to use a truncated passive? In some cases, it allows us to avoid laying blame or taking responsibility:

(9)   Mistakes were made.

The cake was left out in the rain.

In other cases, we may know the result of an action, but the doer is unknown to us:

(10)   The window was broken during the night.

A monument was erected in her honor.

And, in still other cases, the doer is obvious, so expressing it would be redundant:

(11)   The governor's address was delivered at 9:00 P.M.

All my money was lost on a bad gambling bet.

For good or for evil, the truncated passive is a popular construction, used widely even by those who frown upon the use of the passive voice in general.

DISCUSSION EXERCISE 9.5

1. Turn the following active clauses into truncated passives:

Someone left that dog out all night.

A thief stole all my jewels.

I ruined your valuable painting.

Everyone respects my family.

The treasurer gave the treasurer's report at the end of the meeting.

2. Do any of the truncated passives you created seem preferable to the full passive? Why?

3. Turn the following truncated passives into full passives by supplying a doer.

> He was arrested and searched.
>
> The books were read to the children.
>
> Your car got wrecked.
>
> The show was cancelled.
>
> The land got developed.

4. Some of the examples above remind us that the verb *get* can be used as the auxiliary verb in the passive voice instead of *be*. But it also can be an active, transitive verb meaning "acquire" or a linking verb meaning "become." What is *get* in each of the following: auxiliary, transitive, or linking?

> She got a new puppy last night.
>
> Ellen gets red in the face when you compliment her.
>
> The problem got resolved by the lawyers.
>
> They got admitted for free.
>
> Harry got accepted by the army.
>
> Steve got a letter from his uncle.
>
> Everyone got angry.

THE PASSIVE AND STRUCTURAL AMBIGUITY

To review, the elements of the full passive voice are as follows:

Logical Receiver + Auxiliary Verb + Past Participle of the Main Verb + *by* **+ Logical Doer**
Noun Phrase   *Be* or *Get*   **Noun Phrase**

The elements of the truncated passive are as follows:

Logical Receiver + Auxiliary Verb + Past Participle of the Main Verb
Noun Phrase   *Be* or *Get*

All of the sentences of (12) are truncated passives.

> (12)   The gates were opened.
>
>   My complaints were dismissed.
>
>   The objection was overruled.
>
>   The news was expected.

Each has an implied doer and we can reconstruct a full passive or an active sentence by supplying an actual doer. But now consider the sentences of (13):

> (13)   The gates were closed.

The meat was cooked.

My parents were amused all afternoon.

The ice cream was melted.

These sentences can also be construed as truncated passives, and we can supply the implied doer. But there is another structural interpretation of these sentences that looks like this:

Noun Phrase Subject + Linking Verb + Adjective

Under this structural interpretation, there is no implied doer, and the verb phrase is a linking (not an auxiliary) verb followed by an adjective. So, on one reading the sentences of (13) are like those of (14):

(14) The gates were heavy.

The meat was raw.

My parents were angry all afternoon.

The ice cream was sweet.

In other words, the sentences of (13) may be thought of as actions, or they may be thought of as descriptions.

DISCUSSION EXERCISE 9.6

1. What are the two different meanings associated with each sentence of (13)?
2. Why aren't the sentences of (12) structurally ambiguous in the same way that those of (13) are?
3. Construct passive sentences in which the following words are most likely to be interpreted as past participles: *agitated, upset, withdrawn.*
4. Use the same words to construct sentences in which they are more likely to be interpreted as adjectives.
5. Use the word *frozen* in a sentence so that it could be interpreted as a past participle or an adjective.

We have seen in this chapter that the way we form clauses depends in part on the purposes they serve in communication. In the next chapter, we explore in more detail the connection between the form of a clause and its communicative function.

REFLECTIONS

1. In addition to the active and passive voice, there is another uncommonly used voice, sometimes referred to as the *middle voice*. Some examples of the middle voice are the following:

These pants iron easily.

This book reads well.

Why do you think grammarians call this the *middle voice*?

2. In modern English, we cannot freely rearrange the noun phrases of a clause unless we also add the signals of the passive voice to the clause. In Old English, however, it was possible to rearrange the noun phrases of a clause without changing the meaning and without marking it as passive. Consult a history of the English language to find out why it was possible to do this then but not now.

3. The progressive tenses were not used regularly in English until the late eighteenth century, which might explain why they still do not combine readily with the passive voice. Some are more acceptable than others in modern English, as illustrated by Discussion Exercise 9.2. Present and past progressive have acceptable passives: *The house is (was) being built by the contractor.* What about the future progressive? Does it have a corresponding passive? Can you give an example of it?

4. There are some verbs in English that are inherently passive in their meaning, like *receive*. It follows the usual pattern for active and passive voice, but there is no doer associated with *receive*, whether active or passive:

His friend received the package.

The package was received by his friend.

Can you think of other verbs that are like *receive* in this respect?

5. Select about five pages from a textbook or an article in a journal. How many examples of the passive voice do you find, both full and truncated?

6. There is an interesting Supreme Court decision that hinges on the difference between *open*, the adjective, and *opened*, the past participle. It involves the question of whether a van was *open* or *opened*, and whether a search of that van was legal. For the details of this case, read Lawrence M. Solan, *The Language of Judges* (Chicago: The University of Chicago Press, 1993), pp. 4–8.

PRACTICE EXERCISES   (Answers on p. 269)   _____

1. Turn each of the following active sentences into its corresponding full passive, using the verb *to be* as the auxiliary verb.

    1. The realty company sold our house.

    2. The storm caused widespread destruction.

   3. This job requires great sensitivity.

   4. Mixing the two chemicals created an explosion.

   5. A consulting firm is hiring a new chancellor.

   6. Our president will deliver the speech tomorrow.

   7. The librarian left a book at the circulation desk.

   8. Many people have witnessed their cruelty.

   9. The family purchased a week's worth of groceries with your gift.

   10. A retired outfielder threw the first ball.

2. Which of the above have equally acceptable *get*-passives?

3. Name the grammatical functions of the noun phrases in each of the following. Remember that in the passive voice, noun phrases may have more than one function.

   1. The teller handed the customer a roll of bills.

   2. Your proposal has been accepted by management.

   3. The stagecoach was robbed by the bandits.

   4. That unhappy child demands constant attention.

   5. The customer was handed a roll of bills by the teller.

   6. The book was returned to the library by the sheepish patron.

   7. Your party will meet you by the fountain.

   8. Many intelligent people admire good writing.

   9. The suspect was informed of his rights by the police.

   10. Free cheese is distributed by the government.

4. Restructure the following sentence so that the logical indirect object becomes the grammatical subject:

   The doctor gave a clean bill of health to the nervous patient.

5. Make a truncated passive out of each of the following:

   1. One should avoid the passive voice.

2. Someone chopped down that old tree.

3. People expect you to dress appropriately.

4. The legislators passed the bill by a vote of 50–37.

5. The trash collectors pick up the trash on Tuesdays.

6. This TV station broadcasts the news at 6:00 every evening.

7. The opposing team defeated the Yankees.

8. The storekeeper will open the store early tomorrow.

9. The police obtained a warrant to search the house.

10. Everyone prefers laser printers.

6. Why are the full and truncated passives corresponding to the above sentences so close in meaning?

7. Make the following truncated passives full passives by providing a doer:

1. Lunch should be eaten in the cafeteria.

2. The animals were released from their cages.

3. Her cries for help were ignored.

4. The law was repealed.

5. The bell was rung one last time.

6. My car will be repaired.

7. The rules of the competition were rarely understood.

8. Our mail is being sorted right now.

9. Her gardens were planted in May.

10. The students were told to begin the exam.

8. Give the noun phrase in the following sentence that satisfies each description:

The child was given a handful of coins by the clown.

1. grammatical subject: _____

2. logical subject: _____

3. grammatical direct object: _____

4. logical indirect object: _____

5. grammatical object of a preposition: _____

9. Restructure the above sentence so that the logical direct object becomes the grammatical subject.

10. Which of the following are structurally ambiguous? Describe the two possible structures.

   1. The animals were frightened.

   2. His clothes were torn.

   3. The traffic was heavy.

   4. The doors were open.

   5. The doors were closed.

   6. We were amused.

   7. Class was dismissed.

   8. The children were ready.

   9. The movie was exciting.

   10. The dog was tied to a tree.

10

CLAUSE TYPE: DISCOURSE FUNCTION

WHAT IS DISCOURSE FUNCTION?

There are many different reasons for communicating, and English helps us to signal those different reasons by providing us with a range of different clause types. As we saw in the last chapter, we can alter the voice of a clause to signal the relative prominence among the noun phrases. Beyond that, we have different clause types that tell our listeners the purpose we have in addressing them. Those different purposes are called **discourse functions**. One main discourse function is to give information to our listeners (readers). To communicate this purpose, we use a **declarative** clause, as illustrated in (1).

> (1)   The cat ate the dog's food.
>
> Mushrooms grow in damp soil.
>
> The house needs a new roof.

Another function of communicating is to get information from our listeners; for this function, we typically use an **interrogative** clause, as illustrated in (2):

> (2)   Did you cancel your subscription?
>
> Why doesn't she return my calls?
>
> Does the gym have a running track?

A third reason we use language is to get people to behave in certain ways, for which we use what are commonly called commands. The grammatical structure associated with this discourse function is the **imperative**, illustrated in (3).

> (3) Be kind to your sister.
>
>   Give me your phone number.
>
>   Don't guess on the exam.

A fourth discourse function is the **exclamative**, used to express a judgment or a feeling with added emphasis, as in the clauses of (4):

> (4) What a nice person you are!
>
>   How considerate he was!
>
>   How she works!

As you can see, English uses a variety of means to signal these different discourse functions. We might rearrange the order of elements, add a special word, change the form of a word, or leave something out. In writing, we may alter the punctuation; in speaking, we may alter the intonation. The rest of this chapter will explore the particular means we use to signal each of these discourse functions.

Declaratives

The *declarative*, the clause type designed to give information, is the most basic clause type. It is the one most often used in expository writing and probably in ordinary communication as well. Typically, the order of the elements in declarative clauses is subject + verb + object (when there is an object), which is why English is sometimes referred to as an "SVO language." Most of our examples up to this point have been declarative. There is no special marking to indicate that a clause is declarative, except for the period that we put at the end of a declarative sentence in writing. In speaking, the intonation, or pitch, falls at the end.

DISCUSSION EXERCISE 10.1

1. Which of the following are declaratives?
        Don't be a fool.
        Why didn't you stop them?
        I have no time to talk right now.
        Save me a seat.
        He decided to run for office.
        The organization is poorly run.
        Did you see them arrive?

How calmly she received the news.

What an interesting person he is.

The golf course was flooded for days.

2. Can you change the discourse function of the following declarative clauses by changing the order of elements?

You are ready to leave.

He has seen the light.

They didn't like the speech.

3. Now, change the discourse function of the above declarative clauses by changing the intonation.

4. Declaratives can be in the active or the passive voice. Turn each of these active declaratives into their passive equivalents.

A semi hit the bicycle.

The instructor dismissed the class.

Everyone loves mimes.

Her sister has seen her.

The new tenants repaired the sidewalk.

Interrogatives

Yes-No Questions There is much to say about *interrogatives,* those clauses that are designed to get information. There are several different types of interrogatives, because the type of interrogative we use depends on how much we already know and what kind of information we are trying to obtain. For example, suppose I have reason to think that Mary left town this morning, but I need to get verification from someone who knows. I might ask the question in (5):

(5) Did Mary leave town this morning?

In this case, my question has supplied all the relevant information about the event. All I'm asking of my listener is to tell me whether it is true or false. In other words, my listener can satisfy my curiosity merely by saying *yes* or *no.* For this reason, this type of interrogative is called a **yes-no question.**

Yes-no questions are very complicated structures in English, much more so than in many other languages. Let's see what we do to change a declarative clause into its corresponding yes-no question. Consider the following declarative statements and think about how you would turn them into yes-no questions if you needed to find out if they were true or not:

(6) She can do it tomorrow.

I must leave early.

Those girls were laughing at him.

Your computer is making a funny noise.

Dana might leave before dinner.

The children have eaten already.

He had considered the consequences.

Classes will be over on Friday.

You will meet him today.

He would do it for me.

You undoubtedly noticed several things that you must change to create a yes-no question: the intonation, the punctuation, and the order of elements. The falling intonation of the declarative must change to a rising intonation for the question. In writing, we replace the period with a question mark. What about the ordering change? The subject noun phrase and the verb change places. Although these changes do not seem very complicated, Discussion Exercise 10.2 will demonstrate to you that there is more to this question type than first meets the eye.

DISCUSSION EXERCISE 10.2

1. Write the yes-no question that corresponds to each of the following declarative clauses.

    The show starts in five minutes.

    Her father taught her to fish.

    You like to eat strawberries.

    They seemed very nervous.

    The fork goes on the left.

2. How do these differ from the ones you formed from the statements in (6)?

3. What determines whether we simply reverse the subject and the verb or add the verb *do*?

You were able to figure out from these exercises that English yes-no questions are sensitive to whether the statement on which they are based has a helping verb or not. We can give the following rules for phrasing yes-no questions:

If there is a helping verb, reverse the order of the helping verb and the subject noun phrase.

If there is no helping verb, add the auxiliary verb *do* to serve as the helping verb.

How do we know what form of the verb *do* to add? You will notice that in the following sentences the form of *do* varies from question to question:

(7) They left: <u>Did</u> they leave?

Water takes a long time to boil: <u>Does</u> water take a long time to boil?

Those people own that store: <u>Do</u> those people own that store?

The appropriate form of *do* has the same tense and number as the main verb in the declarative. What happens to the main verb in the yes-no question? As you can see, it seems to lose its tense and number and reverts to its base form. Therefore, we can think of the main verb as giving over its tense and number to the auxiliary verb *do*.

DISCUSSION EXERCISE 10.3

1. Expand the above rules for forming yes-no questions by including instructions for choosing the correct form of the verb *do*.

2. Make yes-no questions from the following:

    He has been making a mess.

    She could have been telling the truth.

    They will be taking the first train tonight.

    Which helping verb changes place with the subject if the clause has more than one helping verb?

3. Consider the following yes-no questions:

    Are you happy?

    Is Chicago in Illinois?

    Am I friendly?

    In these sentences, what kind of verb is *be,* helping or main? How do we have to revise our definition of helping verb for the purpose of making yes-no questions?

4. Yes-no questions can be in the active or the passive voice. What is the passive equivalent of each of the following yes-no questions?

    Did the wind destroy the barn?

    Do the students respect the teacher?

    Does that university reject too many students?

    Why doesn't the verb *do* appear in the passive form of the question?

Wh- Questions There is a different kind of question we ask when we are missing a piece of information. In each of the following, for example, we are seeking some specific item of information that we need to complete the thought:

(8) Who is going?

Whom did you call?

What does he want?

Whose did he borrow?

Which do you like?

You will recognize the first words of each as the interrogative pronouns we discussed in Chapter 5. We also mentioned in that chapter that there are other interrogative words that elicit bits of information: *where, when, how,* and *why.* Collectively, these question words are known as *wh- words,* and the type of question they form is called a **wh- question.** Answers to wh- questions, of course, cannot be *yes* or *no.* Rather, the listener is required to supply the missing piece of information as indicated by the interrogative word you have chosen. That missing information might be a noun phrase, in which case you will use *who, whom,* or *whose* if it is human, and *what* or *which* if it is nonhuman.

DISCUSSION EXERCISE 10.4

1. What determines whether we use *who, whom,* or *whose* if the missing informa- tion is a human noun phrase? If you don't remember, consult Chapter 5.

2. What type of information is sought by *where, when, why,* and *how*? Give an ex- ample of each.

3. Some of the interrogative words can also be used as interrogative determiners to ask a wh- question. Which ones can do this? Give examples.

Wh- questions are even more complicated to form than yes-no ques- tions. We have already seen that they require the addition of a special in- terrogative word designed to elicit a particular kind of missing informa- tion. So, one part of the job of asking a wh- question is figuring out which wh- word to use. Another part of the job is deciding where to put it in the question. If you look at the examples of (8) or consider the examples you made up for the exercises above, you will see a consistent pattern: the wh- word always appears at the beginning of the clause. That seems easy enough, but then we notice that sometimes the rest of the clause gets re- arranged when we put an interrogative word in front. Compare the exam- ples of (9) with those of (10):

(9)   Who is blaming us?

      What is making that terrible noise?

(10) Where may I sit?

      How can we help you?

      Why would they remember that?

In (9), the clause retains its SVO order, but in (10), the subject and verb trade places. The difference between (9) and (10) is that in (9), the missing information is the subject; in (10), it is something other than the subject. Two rules we can give for forming wh- questions are the following:

If you are questioning the subject noun phrase, simply insert an interrogative word in the subject position.

If you are questioning something other than the subject, place the interrogative word at the beginning of the clause, then reverse the order of the subject and the verb in the clause.

DISCUSSION EXERCISE 10.5

1. What is the wh- question you would ask to complete the following thoughts?

    (someone?) is knocking at the door

    They are going (sometime?)

    You are saying that (for a reason?)

    Shelly can park her car (somewhere?)

    Paul must respond (somehow?)

2. Which of the above require the subject and verb to switch places? Why?

3. Suppose you wanted to elicit the following information:

    You have been talking to (someone?)

    He was cutting it with (something?)

    They were waiting for (someone?)

    How would you form the questions? Do you have some choice? (You might want to consult Chapter 7.) How is the second rule that we gave above affected by these examples?

The observant reader will realize that we are not finished yet. Notice that the examples above were carefully chosen to avoid an additional problem that becomes evident in the questions in (11):

(11)   They wanted (something?): What did they want?

    The train leaves (sometime?): When does the train leave?

    You do that (somehow?): How do you do that?

    She found the answer (somewhere?): Where did she find the answer?

What you see should be familiar by now. The earlier questions all had helping verbs, and so the order of the subject and helping verb was reversed. But what if the statement on which the question is based has no helping verb? Then we must insert a form of *do* as the helping verb, transfer all the grammatical information from the main verb to *do*, and leave the main verb in its base form. In other words, if we are questioning something other than the subject, we move the interrogative word to the front of the clause and treat the rest of the clause like a yes-no question.

DISCUSSION EXERCISE 10.6

1. Make a wh- question out of each of the following. Which ones require the insertion of the auxiliary verb *do*? Why?

    You have been (somewhere?)

    They traveled to Europe (sometime?)

    Stu is stalling (for a reason?)

    (Someone?) likes to sit in front

    Rebecca managed to do it (somehow?)

    (Someone?) has been using my razor

    The company hired (someone?)

    This is (someone's?) raincoat

    Art turned his back (for a reason?)

    The watchman heard (someone?)

2. Complete the following rules for forming wh- questions in English:

    1. Decide which wh- word to use:

        a. use *who* when the missing information is _____.

        b. use *whom* _____.

        c. use *whose* _____.

        d. use *which* _____.

        e. use *what* _____.

        f. use *where* _____.

        g. use *how* _____.

        h. use *when* _____.

        i. use *why* _____.

    2. If the wh- word is a subject, place it _____.

    3. If the wh- word is not a subject, place it _____.

    4. If the wh- word is preceded by a preposition, _____.

    5. If a nonsubject interrogative word is moved to the front, and the rest of the clause contains a helping verb, _____.

    6. If a nonsubject interrogative word is moved to the front and the rest of the clause does not contain a helping verb, _____.

    7. When *do* is added as an auxiliary verb, it derives its tense and number from

        _____.

    8. When *do* is added as an auxiliary verb, the main verb becomes _____.

3. Give a wh- question to illustrate each step you have completed above.

4. One other difference between oral yes-no and wh- questions is in the intonation. We said earlier that our pitch rises at the end of a yes-no question. What happens to our pitch in a wh- question?

5. It is possible to have more than one wh- interrogative in a question. Suppose you knew the following: *(Someone?) did (something?) to (someone?).* What question would you ask to fill in the missing information?

6. Wh- questions can be in the active or the passive voice. Give the passive equivalents of the following wh- questions:

   What did the fight accomplish?

   When did the storm knock down the power lines?

   How did the manager calm the angry crowd?

   Which did you discover?

   Who brought the chips?

   The last question has several different passive versions. Why?

Tag Questions You will probably agree by now that interrogative formation in English requires significant mental work, although as adult native speakers of English, we do it without conscious effort. But you will notice that people learning English as a second language will have trouble mastering the fine points of questions, as do children who are learning English as their first language. It will be easier to grasp the other types of questions we are about to discuss if you remember some basic principles of question formation in English: it is often sensitive to the difference between helping verbs and main verbs, it often involves some shift of word order, and it typically makes use of some alteration of pitch, or intonation. In writing, the question mark signals an interrogative, but this does not necessarily reflect differences in pitch. For many questions, we also add a word that specifically marks the clause as an interrogative. We have seen that all these features come into play in the formation of yes-no and wh- questions. We see them again in other types of questions.

Another type of question, closely related to the yes-no question both in form and purpose, is illustrated below:

(12)   That woman is your cousin, <u>isn't she</u>?

       Monkeys can't speak, <u>can they</u>?

       The boys haven't returned yet, <u>have they</u>?

       You are listening, <u>aren't you</u>?

The underlined portion of each example above is known as a **tag question**. One purpose of a tag question is the same as that of a yes-no question: to find out if a statement is true or false. Tag questions can have other conversational purposes as well: to get reassurance, to keep a conversation going, or to get someone to admit to something that you believe is true, among others.

1. Finish off each of the following statements with its appropriate tag question:

    It's a nice day.

    You don't love me anymore.

    He's pretty smart.

    They've left already.

    Nancy shouldn't smoke.

2. What is the most likely conversational purpose of each of these tag questions?

We can see similarities in the formation of tag questions and yes-no questions if we compare those of (12) with the following:

(13)   Jan wrote you a letter, didn't she?

         Your children play the piano, don't they?

         That nurse doesn't have any experience, does she?

         The equipment didn't function right, did it?

Both question types are sensitive to the difference between statements with helping verbs and those without. Both require the insertion of the auxiliary *do* as the helping verb when there is no other. Both have the grammatical information from the main verb transferred onto *do*, and both require reverse order of the subject and the helping verb. There are two additional requirements for the tag question:

The subject is a personal pronoun that agrees with the subject of the statement in person, number, and gender.

The tag question reverses the negativity of the statement: affirmative to negative and negative to affirmative.

1. Give the appropriate tag question for each of the following:

    Christmas falls on Monday this year.

    Having money isn't important.

    The movie starts in an hour.

    The meal isn't ready yet.

    You bought a new car.

    Pete didn't see the train.

    Bobbi practices law.

    Computers save us a lot of time.

This car can't run on diesel fuel.

Those pretzels are making him thirsty.

2. There is another kind of tag question that does not reverse the negativity of affirmative statements:

You're insulted, are you?

She left early again, did she?

What is the conversational effect of these tag questions?

3. Complete the following steps that are needed to form a tag question . The completed list should enable someone who has not fully mastered English to predict which tag follows which statement.

1. For the subject of the tag, choose a personal pronoun that matches _____ _____.

2. If the statement has a helping verb, _____ _____.

3. If the statement has no helping verb, _____ _____.

4. If the statement is affirmative, _____ _____.

5. If the statement is negative, _____ _____.

6. The order of elements in the tag is as follows: _____.

4. Not everyone agrees on the tag questions after statements like the following:

There is hardly any sugar left.

He seldom visits us.

You rarely get to see her.

Why do you think we have trouble agreeing on the appropriate tag question? Would studying the rules help us?

5. Why do you think some grammatical purists object to the tag question *aren't I?*, as in *I'm included, aren't I?* In what way does it violate the rule for making tag questions? What are our choices if we don't choose *aren't I?*

6. The following do not conform exactly to the rules that we have developed for tag questions. In what way does each stray from those rules? Why do you think this happens?

Everyone is coming, aren't they?

Someone let the cat out, didn't they?

The baby is cute, isn't it?

There are some left, aren't there?

7. What is the tag question associated with each of the following passive statements?

Selma was arrested by the police.

Selma got arrested by the police.

How are *be* and *get* different with respect to tag questions?

Minor Question Types   There are several other question types that have more limited use than the three we have discussed. One is used in response to a statement and basically echoes the structure of the statement, as in (14):

(14)   I closed down my bank account.
       You closed down your bank account?

       Mary ran off to Australia.
       Mary ran off to Australia?

For this reason, they are called **echo questions**. (Some grammarians call them **declarative questions**.) The intonation and punctuation is that of a yes-no question, but the structure remains that of a declarative statement. Questions like this often signal surprise or disbelief rather than a true interest in getting information. It is also posssible to form an echo question containing a wh- word, as illustrated in (15):

(15)   I faxed the letter to the president of the company.
       You faxed the letter to whom?

       We're going to live in Texas.
       You're going to live where?

Again, you see that the basic structure of the declarative remains intact, and we use a wh- word to register what it is that surprises us. The questions of (14) are called **yes-no echo questions**; those of (15) are **wh- echo questions**.

DISCUSSION EXERCISE 10.9

1. Form an echo question as a response to the following statements. You will have a range of options because it is possible to register disbelief about the whole statement (yes-no echo) or about individual parts of it (wh- echo).

   I'm moving to Canada with my accountant.

   She bought a Jaguar to keep in her living room.

2. It is also possible to register disbelief or surprise about several different parts of a statement at once. Try it with the following statement:

   Sandi is going to the opera with Placido Domingo.

Another minor question type is not recognized as a question by everyone. It is embedded inside a sentence and is used when we have questions like the following in mind:

(16) I wonder (did she go?)
   She doesn't know (did she pass the exam?)
   They asked (could they ride with us?)

   I wonder (what did she say?)
   She doesn't know (how did she do that?)
   They asked (where were we staying?)

Although some dialects of English express these questions just as they appear in (16), formal standard English requires that they resume their declarative structure, inserting *if* or *whether* before a yes-no question and a wh- word before a wh- question. These structures, as they appear underlined in (17), are called **embedded questions**.

(17) I wonder <u>if she went</u>.
   She doesn't know <u>whether she passed the exam</u>.
   They asked <u>if they could ride with us</u>.

   I wonder <u>what she said</u>.
   She doesn't know <u>how she did that</u>.
   They asked <u>where we were staying</u>.

DISCUSSION EXERCISE 10.10

1. Embed each of the following questions into a larger sentence, changing the form as required by standard English.
   Why did they change the date?
   Are they still friends?
   When will this job be finished?
   Am I still in the running?
   Did the class meet yesterday?
2. Name the question type in each of the following:
   Who is calling my name?
   You're always right, aren't you?
   Are you having a good time?
   She wonders if the paint will cover the stains.

He hasn't studied for the exam?

You told her what?

Imperatives

Imperatives are those clause types designed to get people to behave in certain ways. They don't necessarily require a verbal response, but we hope for some action or some change in mental state when we utter them. Because they are designed to give orders directly to the person we are addressing, the grammatical subject is always *you* (singular or plural), but the most noticeable characteristic of imperative clauses is that they do not require the grammatical subject to be expressed at all, giving us the notorious "understood *you*." The following pairs are equivalent in intent, although they may differ slightly in force or emphasis:

(18)   You leave this instant!

Leave this instant!

You bring me a cup of coffee!

Bring me a cup of coffee!

As indicated by the examples, in written English we often punctuate imperatives with exclamation points.

DISCUSSION EXERCISE 10.11

1. Create an imperative clause with each of the following verbs: *read, stay, speak, jump.*
2. Which of the Discussion Exercises in this chapter use the imperative to get you to do something?
3. The verb in the imperative looks like it could be the present tense or the base form, since they are often identical. But consider the verb *to be* in its imperative form. Does that help you decide which it is?
4. Imperatives with the *you* subject expressed could also be analyzed as declarative statements. Give the two possible interpretations of *You plan to be out of here by tonight.*
5. Imperatives do not readily combine with the passive voice, although it is possible. Can you think of an example? (Don't be fooled by the question!)

Exclamatives

Exclamatives are a minor clause type that, as we said earlier, allow us to express a judgment or feeling with added emphasis. You'll notice that they use some of the same signals as interrogatives: the addition of *how* or *what*, and rearrangement of word order. In writing, they are often punctuated with an exclamation point. Below are some additional examples of exclamatives.

(19) What a fine job you've done.
      What an intricate pattern she's woven.
      What a silly idea that is.

(20) How ordinary he seems.
      How selfish you've become.
      How petty they are.

(21) How carefully they've scrutinized the report.
      How intensely they love each other.
      How neatly you write.

(22) How she trembles at the sight of him.
      How they studied before that exam.
      How they laughed that night.

What are the things we can exclaim about, grammatically speaking? In (19), it is a noun phrase; in (20), an adjective; in (21), an adverb; and in (22), a verb phrase. For noun phrases, we use *what* and for all the others we use *how*.

DISCUSSION EXERCISE 10.12

1. You'll notice that in three of the four sets of examples above, we bring forward the thing we are exclaiming about. In which set does the word order remain the same as for declaratives?

2. Turn each of these into an exclamative to emphasize the underlined element.
      They dance <u>well</u> together.
      The sky looked <u>beautiful</u> that night.
      This is <u>a disorganized mess</u>.
      I <u>envy you</u>.
      He is <u>a fine physician</u>.

3. We are restricted in the kind of noun phrase we can exclaim about in this fashion. What is the requirement on the determiner?

4. Some people think exclamatives sound old-fashioned. Others think they mark speech specifically addressed to children. What is your own opinion about the use of exclamatives?

CROSSOVER FUNCTIONS OF CLAUSE TYPES ·

We have seen throughout this chapter that there are certain clause types designed for certain discourse functions. When we choose one of these with its characteristic markings, we signal to our listener what our purpose is in communicating at that moment. But, as we have seen before, categories in English tend to be slippery and there is a tendency for elements of one category to shift into another. Such is the case for clause types. We find in normal everyday communication that clause types designed for one purpose can sometimes fulfill another. For example, consider the clauses of (23):

> (23)   What do you think you're doing?
>
> Do you think I'm stupid?
>
> Where are your brains?

As we can see, these all have the form of an interrogative, and they may be interpreted literally: *I think I'm teaching grammar; Yes, I think you're stupid; My brains are in a jar on my desk* (said by one anatomy student to another). But, as speakers of English, we know that these can also have the force of a declarative statement that means something different from the literal meaning: *You're doing something wrong; You're insulting my intelligence; You've done something stupid.* As declaratives, they do not require answers, and attempts to answer the literal question may be seen as inappropriate. Questions such as these are often called *rhetorical questions.*

Can we ever do the reverse, use a declarative statement to get information? Consider the following:

> (24)   I have to know your name.
>
> I need to know the capital of France.

Although these are literally declarative statements reporting on a need of the speaker, it would not be conversationally inappropriate to interpret them as requests for information and provide the answers: *It's Fran; It's Paris.*

The most interesting of the crossover functions of clauses involve the imperative. Although there is a specific clause type designated for giving commands, use of that clause type is often viewed as too blunt or too aggressive. Direct imperatives work well under urgent circumstances:

> (25)   Watch out!
>
> Duck!
>
> Follow that car!

But under normal, non-urgent circumstances we look for ways to soften the effect of the imperative. We may simply add the word *please*, but that is not the only strategy we employ. Often what we do instead is use another clause type. For example, we might use a declarative statement that expresses our own needs rather than issue a direct command:

(26) I wish you would stop talking.
       I need you to read those files by tomorrow.
       I expect you to be on time.
       I'd like you to call your grandmother.

Or we might use the interrogative with a modal verb:

(27) Can you help me?
       Would you close the door?
       Could you wait here a moment?
       Will you give me a pencil?
       Must you talk so loud?

Or we might use a declarative with a modal verb:

(28) You might want to check your oil more often.
       You could be a little nicer.
       You will arrive on time from now on.

All of these, of course, have literal interpretations, but they are not the ones that first come to mind. The softened imperative is the more natural interpretation: an action is a more appropriate response than the receiving or giving of information.

DISCUSSION EXERCISE 10.13

1. Give a literal answer to each of the following. Under what circumstances might they be considered rhetorical?

    Do you have two hands?

    Am I your maid?

    Can't you see that I'm busy?

    Where is your sense of decency?

    Why don't you stop talking?

2. Give a softened version of each of the following imperatives. Under what circumstances might the direct imperative be used?

    Lend me fifty dollars.

    Shut the window.

    Write me a letter of recommendation.

Hand over that gun.

Don't vote for that amendment.

3. Can you find Discussion Exercises in this chapter that tell you to do things by means other than the direct imperative?

Clause type, as we have seen in the last two chapters, is determined by voice—active versus passive—and by discourse function. There is another factor that determines clause type: affirmative versus negative. This will be the focus of the next chapter.

REFLECTIONS

1. There are two ways of asking someone for the time:

   Do you have the time?

   Have you the time?

   What does it tell you about how the verb *have* is treated in yes-no questions? Does one sound more natural to you than the other?

2. You might hear a child say *Did he be happy?* What part of the rules for yes-no questions has she not yet figured out ?

3. You might hear a child say wh- questions like *What daddy can do?* or *Where mommy put the ball?* Which part of wh- question formation has he not mastered yet? If you have regular access to a child learning English, record the child's questions over a period of a few days and then analyze them to see which rules have been mastered and which haven't.

4. Consider these questions, which appear in Shakespeare's *Julius Caesar*:

   What mean'st thou by that?

   What means this shouting?

   Why stare you so?

   Think you to walk forth?

   Comes his army on?

   In what respect has question formation changed since Shakespeare's time?

5. There are two possible tag questions for this statement: *He has a lot of money.* What are the two? What does it tell you about the classification of the verb *have* for the purposes of tag-question formation?

6. One of the amazing facts about tag questions is that all children learning English go through the trouble to learn the rules for making them

even though we have one word that can substitute for all of them, similar to the French *ne c'est pas*? What is that word?

7. There are two other structures that are also considered to be imperatives of sorts. These are illustrated below:

    Somebody get the phone.

    Let each man fend for himself.

    Let's go to the movies.

    In what ways are these different from the imperatives we have discussed? Why do you think they are called imperatives?

8. The verb *look* often precedes the exclamative without changing the intent of the utterance: *Look what a nice job you've done!* Can you think of another verb that acts like *look* in this respect?

9. It is interesting to think about how we learn that some statements are not to be taken literally, especially softened imperatives. If you have the opportunity, you might want to observe how adults give orders to children. (Is there an imperative in this paragraph?)

10. Each discourse function described in this chapter has a specific sentence type associated with it. There are many other minor discourse functions that are signaled by the use of a verb that explicitly tells the function of the utterance. Some examples are *I promise not to talk* and *I warn you not to lie.* You'll notice that the first statement functions as a promise, and the second functions as a warning. What other discourse functions are signaled in this way? (Hint: some of the more formal ones are often preceded by the word *hereby*.)

PRACTICE EXERCISES   (Answers on p. 271)   _____

1. Name the clause-type in each of the following: *declarative, interrogative, imperative,* or *exclamative*; provide the appropriate punctuation at the end.

    1. What an interesting course this is

    2. How can I help you

    3. Leave me alone

    4. She's thinking of buying a hamster

    5. Let's settle our differences

6. How easily you judge others

7. Does rice have a lot of calories

8. When did the race begin

9. That teacher taught me how to think

10. Everyone remain calm

2. Create a yes-no question corresponding to each of the following declarative statements:

1. We need to rotate the tires.

2. Help is on the way.

3. You can give me an estimate on the costs.

4. He did the assignment before class.

5. The laborers rested after lunch.

6. Word-processing saves us a lot of time.

7. We must observe the rules.

8. They are aware of the danger.

9. I spelled the word wrong.

10. This dog won't fetch.

3. Formulate the standard English wh- question that will elicit the missing information in each of the following:

1. You borrowed (someone's?) book

2. (Someone?) needs a ride

3. They can do (something?) for me

4. I can find good corned beef (somewhere?)

5. She was asking for (someone?)

6. (Something?) fell on my head

7. She ordered (one of several?)

    8. He remained silent (for a reason?)

    9. You wish to speak with (someone?)

    10. The scandal destroyed (someone?)

4. Which of the above questions required the insertion of the auxiliary verb *do*? Why didn't the others require *do*-insertion?

5. Form the appropriate tag question for each statement.

    1. The tornado hit the center of town.

    2. Bees communicate by dancing.

    3. We can't respond to that inquiry.

    4. Medical school is difficult.

    5. The children had a good time.

    6. You went sailing yesterday.

    7. The plants have all died.

    8. I have ink on my face.

    9. Stanley has lost the tickets.

    10. Barbara looks good.

6. Which tag questions in the preceding exercise required the insertion of the auxiliary *do*? Why?

7. What makes it awkward to follow the rules for tag-question formation for each of the following statements?

    1. She can barely read it.

    2. No one likes housework.

    3. Someone called here yesterday.

    4. That baby is smart.

    5. They hardly ever get to see each other.

8. Name the question type in each of the following:

   1. You haven't done it yet?

   2. I wonder if I need a raincoat.

   3. What's for dinner?

   4. You're upset, aren't you?

   5. To whom shall I direct your call?

   6. We're having dinner with whom?

   7. Can you spare a dime?

   8. Kenny didn't show up, did he?

   9. Tell me what you saw.

   10. Why can't we stay?

9. Turn each of these into an exlamative clause that emphasizes the underlined element.

   1. Al is <u>a wonderful parent</u>.

   2. That building is <u>ugly</u>.

   3. My heart <u>aches for you</u>.

   4. They have <u>a cute baby</u>.

   5. She complained <u>bitterly</u>.

   6. This is <u>awful</u> for you.

   7. He <u>boasts about his children</u>.

   8. Senta performed <u>well</u> in the recital.

   9. Kara and Jan were <u>proud</u> of her.

   10. This is <u>a tiresome discussion</u>.

10. Give two softened versions of each of the following imperatives, one using a declarative and one using an interrogative:

   1. Take the dog for a walk.

   2. Pick up some milk on your way home.

3. Talk to me about your concerns.

4. Help me with these groceries.

5. Clean up this mess.

6. Look for a job.

7. Rethink your demands.

8. Host a reception for them.

9. Consider your alternatives.

10. Get there on time.

11

CLAUSE TYPE: AFFIRMATIVE VERSUS NEGATIVE

WHAT IS NEGATIVITY IN GRAMMAR?

In addition to having voice and discourse function, clauses in English are generally marked as being either **affirmative** or **negative**. The *affirmative,* or positive, clause has no special marking, but the *negative*—that which negates, or expresses a "no" answer—is marked in a variety of ways. Being able to say "no" is a very important feature of communication, so language provides us with many different ways to get this meaning across. As you can imagine, it would be highly inconvenient for us if the grammatical markings for negation were so inconspicuous that people would be inclined to miss it in our conversation. English grammar allows us to make it clear that we *don't* want to buy swampland or cemetery plots, even if there is static on the telephone; that the new VCR we are returning *doesn't* record accurately; that we *can't* serve on still another committee; and that the report *won't* be ready until tomorrow. No one likes to sound negative, but being able to express negativity is essential to our well-being.

VERB NEGATION

Negation in English is expressed by attaching it to different lexical categories. One of these categories is the verb. You can see how this works in the clauses of (1):

(1) You may not leave yet.
     The letter has not arrived.
     The secretary could not reach the client.

A child cannot sit still for that long.

The door must not be opened.

I shall not be here tomorrow.

Nancy is not coming to the party.

They should not allow that to happen.

The class will not meet tomorrow.

The train might not be on time.

So far, there appears to be a very simple procedure for **verb negation**: add the word *not* after the verb.

DISCUSSION EXERCISE 11.1

1. As speakers of English, you know that in less formal usage, a contraction of the verb and *not* can occur. Which of the sentences in (1) have corresponding contractions? Which contractions are irregular?
2. Express the following with a contraction:

   I am not allowed to smoke.

   How is this different from the contractions you made for Exercise 11.1.1?
3. The notoriously nonstandard *ain't* is a contraction of a verb and *not*. What, specifically, is it a contraction of in each of the following?

   I ain't going.

   She ain't happy.

   They ain't laughing.

   We ain't seen it.

   He ain't left yet.
4. Why can't the following affirmative statements be made negative by the simple insertion of *not* after the verb?

   The ship sails at noon.

   My parents speak Spanish.

   The dog ate its food.

If you were able to answer Discussion Exercise 11.1.4, then you know that verb negation is sensitive to the difference between helping and main verbs, a distinction you are familiar with by now. Just as is the case for many interrogatives, verb negation requires the insertion of the auxiliary verb *do* if a main verb is negated (with the verb *be* treated as helping regardless of whether it is actually main or helping). If we negate the verbs in Exercise 11.1.4, the results are as follows:

(2)   The ship does not sail at noon.

My parents do not speak Spanish.

The dog did not eat its food.

We may also use the contractions *doesn't, don't,* and *didn't,* respectively. The form of *do* is dependent on the tense and number of the verb in the affirmative, and once the affirmative verb hands over its information to *do,* it reverts to its base form. This probably sounds more orderly than it did when we first came across this arrangement for interrogatives. Although there is a complicated set of adjustments involved in making verbs negative, the process is essentially the same as that used to turn declaratives into interrogatives.

DISCUSSION EXERCISE 11.2

1. Make the following clauses negative by negating the verb:

   I might see that film.

   The answers are in the book.

   The club has met recently.

   They are enjoying the course.

   My boss will let me have Saturday off.

   Jane sees the robin.

   Steve cooked dinner last night.

   She withdrew her funds in time.

   The rebels won the battle.

   Aerobics makes her energetic.

2. Complete the following rules for negating clauses by negating the verb:

   1. If there is no helping verb, insert _____

   2. *Do* derives its tense and number from _____

   3. The main verb becomes _____

   4. Insert *not* after _____

NEGATION OF INDEFINITES

Another important way we have of expressing negation is to use the negative form of an indefinite word that begins with *some-*. We call this **indefinite negation**. The affirmative statements in (3) can be made negative by using the corresponding negative form of the indefinite, as illustrated.

(3) I need <u>something</u>.   I need <u>nothing</u>.

     She went <u>somewhere</u>.   She went <u>nowhere</u>.

They went <u>sometime</u>.   They <u>never</u> went.

He saw <u>someone</u>.   He saw <u>no one</u>.

In many instances, you can get the same meaning across by negating the verb, but if you negate the verb, standard English requires that the indefinite word begin with *any-*. (But see Reflections 7 at the end of this chapter.)

(4) I do not need <u>anything</u>.

     She did not go <u>anywhere</u>.

     They did not go <u>anytime</u>.

     He did not see <u>anyone</u>.

DISCUSSION EXERCISE 11.3

1. Make the following clauses negative by using a negative indefinite:

    I have somewhere to go.

    You can visit me sometime.

    Somebody showed up.

    He loves someone.

    Something made me nervous.

2. Make the same clauses negative by negating the verb and using an *any*-indefinite. Which of the above cannot be made negative in this way?

3. Here is another grammatical connection between interrogatives and negatives. Turn the following into yes-no questions. What is the similarity between these and negation? How are they different?

    He saw something.

    Someone came.

    This is leading somewhere.

4. The word *ever* is another affirmative form of *never*. How is its use different from *anytime*?

A restriction on standard English negation in modern times is that we may not have two negatives in one clause, the so-called *double negation*. As we discussed in Chapter 1, this is part of the legacy from the eighteenth-century grammarians, who reasoned that two negatives would cancel each other out and become an affirmative. Since the eighteenth century, clauses like the following have been considered nonstandard:

(5) He didn't see nothing.

     I'm not going nowhere.

     They don't need nothing.

Utterances such as those in (5) are common in many dialects of English and are understood by everyone to be negatives, not affirmatives. They do not mean he sees something, or I'm going somewhere, or they need something. In ordinary language use, negatives reinforce one another rather than cancel one another out, regardless of the rule that says otherwise.

DISCUSSION EXERCISE 11.4

1. There is one interpretation of the sentences of (5) that is affirmative, where the negatives do cancel each other out. Can you put them in a context in which they would have affirmative meaning? How do we signal that we want the sentence to have the affirmative interpretation?

2. Are the following examples of multiple negation considered to be standard English or not?

    I don't never have no money.

    We don't never travel nowhere.

    According to the eighteenth-century grammarians' reasoning, should they be standard?

3. Although not part of standard English, negations like the following are used in some dialects of English:

    Can't nobody help me now.

    Don't nobody move!

    What are the rules for forming these negations? In what ways are they different from the standard English rules?

NOUN NEGATION

Nouns are another lexical category that can be negated. **Noun negation** is accomplished by using the determiner *no*. You'll notice in the pairs of sentences in (6) that some verb negations can also be expressed as noun negations:

(6) I don't have any time.
    I have no time.

    There isn't any milk left.
    There is no milk left.

In the second of each pair, *no* works as a determiner in the same way that *the* and *some* do. (*Some* changes to *any* when the verb is negated.)

DISCUSSION EXERCISE 11.5

1. Turn each of the following into an equivalent statement by negating a noun:

    There isn't any point to this conversation.

    I do not see any reason to continue.

She does not expect any compensation.

This restaurant does not serve any liquor.

That teacher doesn't have any patience.

2. Can verb negation and noun negation be used in the same clause in standard English? Give a sentence that illustrates the two used together.

ADJECTIVE AND ADVERB NEGATION

Still another way that we can express negation in English is by using a negative prefix with an adjective or an adverb, called **adjective** and **adverb negation**. The following pairs illustrate this alternative to verb negation:

(7) They are not lucky.
     They are unlucky.

     She is not satisfied with the results.
     She is dissatisfied with the results.

     He did not speak truthfully.
     He spoke untruthfully.

     You do not write legibly.
     You write illegibly.

There are many different negative prefixes in English and, for the most part, we cannot predict which adjective or adverb will be assigned which prefix. Probably the most common is *un-*, and it is often the one that people use instead of the less common ones required by standard English. One such common nonstandard usage is *unattentive* for *inattentive*.

One special feature of adjective and adverb negation is that it may occur in standard English together with verb negation, and when the two occur together they have the effect of canceling each other out, just as the eighteenth-century grammarians said they would. Consider the examples of (8):

(8) I am <u>not unhappy</u> with the results.
     She spoke <u>not insincerely</u>.
     They are <u>not intolerant</u>.

Although we might quarrel about the exact meanings (see Reflections 4 at the end of this chapter), the sentences of (8) are roughly the equivalents of those in (9):

(9) I am happy with the results.

She spoke sincerely.

They are tolerant.

DISCUSSION EXERCISE 11.6

1. Make each of the following adjectives negative by adding a prefix. Is there disagreement on any of these? *regular, secure, possible, courteous, legal, imaginable, violent.*

2. Although there is very little predictability in these prefixes, there is some. For example, can you think of a reason that the following adjective and adverb roots all take the prefix *im-* as opposed to *in-? mobile, maturely, material, politely, pious, probably?*

3. What is the requirement for the roots that permit the prefix *ir-?* What about *il-?*

4. Express each of the following as a combination of verb negation and adjective negation:

They are honest.

His attention is flattering.

These facts are relevant.

NEGATION OF COMPOUNDS

Another way we can express the negative is by **compound negation**. Compounds of constituents can be negated with the correlative coordinating conjunction *neither . . . nor.* Many different constituent types can be negated in this way:

(10)   We are neither sorry nor ashamed. (adjectives)

They neither sang nor danced last night. (verbs)

She neither helped the victim nor called the police. (verb phrases)

I'll have neither the pie nor the ice cream. (noun phrases)

You spoke neither convincingly nor eloquently. (adverbs)

He is neither for nor against the proposal. (prepositions)

We expected neither him nor her. (pronouns)

When compound clauses are negated in this way, a curious thing happens, as you can see in the sentences of (11):

(11)   Neither can he speak nor can he walk.

Neither will they read the books nor will they return them.

Neither have you answered my question nor have you proven me wrong.

You noticed, of course, that the subject and verb of the clauses change places, and you also undoubtedly noticed that the verbs are helping verbs. What do you suppose happens if there is no helping verb? The sentences of (12) illustrate what you were surely able to figure out:

(12)   Neither did he speak nor did he walk.

Neither do they read the books nor do they return them.

Neither did you answer my question nor did you prove me wrong.

Here is still another instance in which English grammar is sensitive to the distinction between the presence or absence of a helping verb, requiring the insertion of *do* when no helping verb is present.

It is also worth pointing out here that *nor* can function independently of *neither*, just as *or* can function independently of *either*. Most commonly, it occurs in the second of two compound clauses when the first is negated in some way, as illustrated below:

(13)   We can't admit the truth, nor can we face the shame.

The child is not happy, nor is she healthy.

I never sleepwalk, nor do I have nightmares.

She expects nothing, nor does she get anything.

DISCUSSION EXERCISE 11.7

1. Give an example of a *neither . . . nor* compound for each of the following constituent types: noun phrase, verb phrase, adjective, pronoun, clause. Do all these constructions occur in your everyday spoken English? Do any seem especially formal?

2. *Neither* can function alone without *nor*. What does it mean in the sentence *I want neither*? What category of negative would you put it under?

3. Why do you think *anything* (as opposed to *nothing*) is required in the second clause of the last sentence of (13)?

PARTIAL NEGATION

One form of negation that we have mentioned before is what might be called **partial negation**. This is achieved through the use of the partially negative adverbs *seldom, rarely, barely, scarcely*, and *hardly*. You will remember from our discussion of tag questions in Chapter 10 that these adverbs lend some negativity to a clause but do not entirely negate it. That is why there is some variability in the tag questions that follow them, since forming a tag question requires that we reverse the negativity of the statement it follows. Notice again how odd either tag sounds in the examples of (14):

(14)   You hardly spoke to him, didn't you?
                                      did you?

       She seldom gets out, doesn't she?
                                      does she?

There is another interesting feature of these partially negative adverbs. Because they are adverbs, there is some flexibility in their placement in sentences. In addition to following the subject, as in (14), some of them may also precede the subject. But notice what happens when they precede the subject:

(15)   Rarely can I find good pasta sauce.
       Seldom will you encounter true friends.
       Scarcely had the lecture begun, when he got up and left.

(16)   Seldom does one hear pleasant news.
       Just barely do I get my work finished.
       Rarely did she get a raise in those days.

Yet again we see a grammatical situation in which what you do depends on whether the sentence has a helping verb or not. If there is a helping verb, we reverse the subject and verb. If there is no helping verb, we add a form of *do*, taking all the information from the main verb and leaving the main verb in its base form. You may never say sentences like those of (15) and (16), but by now you can probably describe their formation with speed and eloquence!

DISCUSSION EXERCISE 11.8

1. Which of these have a corresponding sentence with the adverb before the subject?

       They hardly ever visit us.

       We can barely hear you.

       You seldom long for company.

       He scarcely made a living.

       She rarely does her homework.

2. These same placement rules operate for some other adverbs as well. Show how the adverb *often* can appear before or after the subject. What adjustments are necessary to the sentence if it appears before the subject? Can you think of other adverbs that behave this way?

3. What are the yes-no questions corresponding to the statements in Exercise 11.8.1.? Do you think they sound odd? Why? How would you answer them?

4. Negativity is independent of voice and discourse function. Clauses have all three, so we might refer to the following clause as an *active declarative affirmative* clause:

   The police arrested the thief.

   For purposes of grammatical description, this type is often considered the most basic clause type, from which the others are derived. What clause type is each of the following?

   The police did not arrest the thief.

   The thief was arrested by the police.

   The thief was not arrested by the police.

   Did the police arrest the thief?

   Didn't the police arrest the thief?

   Was the thief arrested by the police?

   Wasn't the thief arrested by the police?

   Arrest the thief!

   Don't arrest the thief!

Until this point, our discussion of English grammar has focused on clauses and their components, or constituents. For the most part, our examples have been single clauses, which we have sometimes called *sentences*. What we will see in the next two chapters is that clauses and sentences are not the same thing. The sentence is a larger unit of language organization that is made up of clauses. A sentence might be made up of just one clause, and so it was not wrong of us to refer to our sample clauses as sentences. But sentences are often made up of more than one clause. Our next task is to learn how clauses can be arranged to make up sentences, which will complete our study of English grammar from its lowest to its highest levels of organization.

REFLECTIONS

1. Why do you think so many people use the contraction *ain't* even though everyone knows that it is nonstandard? Can you think of any popular songs that use *ain't* in the lyrics? What purpose does it serve?

2. Consider the following from Chaucer's *Man of Law's Tale*:

   And therefore he, of ful avysement

   Nolde (would not) *nevere write in none of his sermons.* . . .

   How did negation in Chaucer's time (fourteenth century) differ from modern English negation? [See *The Complete Poetry and Prose of Geoffrey*

Chaucer, 2nd ed., ed., John H. Fisher. (New York: Holt, Rinehart and Winston, 1989), p. 84, Instructor's Edition.]

3. Some languages typically use more than one word at a time to signal negation, such as French and Spanish, as illustrated below:

   French: *Je ne sais pas*

              I-not-know-not

   Spanish: *Yo no sé nada*

              I-not-know-nothing

   Do you know any other languages that use more than one negative per clause?

4. It can be argued that verb negation and adjective negation do not produce exactly the same meaning. For example, the two sentences below are not necessarily equivalent in meaning:

   He is not lucky.

   He is unlucky.

   What is the difference between them? Similarly it can be argued that the following are not strictly equivalent:

   He is not unlucky.

   He is lucky.

   What is the difference between these? It is interesting to note that these differences are associated only with gradable adjectives. For nongradable adjectives, the affirmative and the doubly negated one mean the same. For example, *reversible* and *not irreversible* have the same meaning.

5. It says in the Declaration of Independence that "all men are created equal, that they are endowed by their Creator with certain unalienable rights." In what way is this different from modern English usage?

6. Here is some dictionary work:

   1. In what sense is *irregardless* a double negative? Is it considered to be standard English?
   2. What is the function of the prefix *in-* in *inflammable*?

7. In this chapter we said that *some-* words must change to *any-* words when the verb is negated. Strictly speaking, it is possible to leave the *some-* words in the negative, but there is a slight difference of meaning. What is the difference in meaning between these two sentences?

   I don't have any money.

   I don't have some money.

8. These are some negative statements from Shakespeare's *Julius Caesar*: How are they different from modern English negation?

Knew you not Pompey?

Forget not in your speed.

Fear him not, Caesar.

I know not what you mean by that.

Review Reflections 4 at the end of Chapter 10. Can you make a general statement about questions and negatives in Shakespeare's time, compared to those of today?

PRACTICE EXERCISES   (Answers on p. 273)   _____

1. Express each of the following using a contraction. Which are irregular? Which would you never say?

    1. You should not speak ill of the dead.

    2. Karel cannot cope with the situation.

    3. Bill may not play today.

    4. Sal is not listening to you.

    5. We were not expecting you.

    6. I am not surprised.

    7. We shall not be daunted.

    8. They must not think we are ungrateful.

    9. He will not be allowed to perform.

    10. That would not help me.

2. Which of the following statements require the insertion of *do* in the negative? Why?

    1. The games have begun.

    2. She needs to consider the alternatives.

    3. The milk is sour.

    4. We could meet you after class.

    5. I do my chores in the evening.

    6. Reading puts me to sleep.

   7. Expect the worst.

   8. The felon was acquitted.

   9. These pretzels are making me thirsty.

   10. I ordered the cake for the party.

3. Express the negative of each of these in two different ways:

   1. The printer has ink.

   2. A reasonable person would tolerate this behavior.

   3. I have an idea.

   4. She received compensation for the job.

   5. He got love from his grandparents.

4. Which of the following conform to the rules of standard English negation and which are nonstandard?

   1. He doesn't eat neither meat nor chicken.

   2. I never did see no thief.

   3. Nothing can be done about it.

   4. I can tolerate no alcohol.

   5. We didn't expect no visitors.

   6. They are neither tired or hungry.

   7. Can't nobody blame him for his actions.

   8. That doctor doesn't give no free samples to nobody.

   9. She neither cooks nor sews.

   10. My handwriting is not illegible.

5. Give an equivalent negative statement for each of the following by negating a noun:

   1. I don't see any people.

   2. Doesn't any proposal suit you?

   3. She won't accept any cash.

4. There isn't any reason to stay.

5. Why weren't there any police at the event?

6. Give a one-word negative for each of the following adjectives or adverbs: *happily, decent, proportionately, advisable, politely, relevant, literate, savory, toxic, remarkably.*

7. Perform ellipsis on the following negative clause compounds so that what remains is a negative compound of a lower-level constituent:

1. We are not proud and we are not arrogant.

2. I didn't clean the garage and I didn't sweep the driveway.

3. She didn't speak clearly and she didn't speak accurately.

4. They won't tolerate your laziness and they won't tolerate your impudence.

5. Smoking is not good for you and drinking is not good for you.

8. Make each of the following statements partially negative by using a partially negative adverb. For which can the adverb be placed in front of the subject?

1. I go to the movies during the week.

2. Marie studies with Ken.

3. We arrived on time.

4. The farmers were able to support their families.

5. He remembers their meeting.

6. The patient could speak.

7. There is enough time for this exam.

8. The tenants complained to the landlord.

9. The beams held up the roof.

10. There is some food in the refrigerator.

9. Make each of the following negative in at least two different ways:

1. This food is edible.

 2. We went somewhere last night.

 3. Our approach is confrontational.

 4. There is some reason to worry.

 5. These reviews are spectacular.

 6. She bought some bonds.

 7. I'll read it sometime.

 8. There is paint in the garage.

 9. A car can travel this road.

 10. That star has fan clubs.

10. Describe each of the following sentences according to its voice, negativity, and discourse function:

 1. Did you lose your keys in the parking lot?

 2. The broadcast was interrupted by the storm.

 3. The students weren't permitted to talk during exams.

 4. Don't be misled by false advertising.

 5. Did the soup taste salty?

 6. You didn't understand my question.

 7. Arrive on time for your next appointment.

 8. Can't the purchase be returned by the customer?

 9. I am uneasy about the truce.

 10. Don't call me before ten o'clock.

12

COMBINING CLAUSES INTO SENTENCES

HOW IS A SENTENCE DIFFERENT FROM A CLAUSE?

It is now time to take a closer look at the highest level of grammatical organization, the **sentence**. As we said at the end of Chapter 11, sentences are made up of combinations of **clauses**, those constituents that are themselves made up of a noun phrase and a verb phrase. Sentences consist of one or more clauses that bear certain relationships to one another. The simplest kind of sentence is the one made up of only one clause. You may have noticed that we used the terms *clause* and *sentence* more or less interchangeably in the preceding chapters, since most of the examples we used were one-clause sentences. In this chapter, we will be more careful to distinguish between the two so that we may accurately describe the processes of sentence building.

SENTENCE BUILDING THROUGH COORDINATION

We have already talked about how we build sentences by combining clauses through **coordination** (or compounding). You will recall that in this process we use the coordinating conjunctions, either simple or correlative, to link two or more clauses together. The relationship between, or among, the clauses is determined by the choice of coordinating conjunction. For example, if *and* is used, it may simply link two facts together in a neutral way, as in (1):

(1)   Chicago is in Illinois and Utica is in New York.

Or, it may suggest that the events happened in the order that they appear, as in (2):

(2)   She graduated and she got a job at the local bank.

Or, it may suggest a causal relationship, as in (3):

(3)   Do that again and you're history.

So allows the first clause to give a reason for the second:

(4)   He was exhausted, so he went to bed.

But and *yet* imply that the second clause is contrary to the expectations of the first:

(5)   Bob worked very hard, but he didn't get a Christmas bonus.
      The plane had been thoroughly inspected, yet it crashed.

Or, as you know, creates a disjunction, presenting the clauses as alternatives:

(6)   You should put your money in the bank or you should invest it.
      Speak now or forever hold your peace.

For allows the second clause to offer a reason for the first:

(7)   He returned the wallet, for he was an honest man.

Nor follows a negative clause and means "and not":

(8)   The water is not safe to drink, nor is the food edible.

DISCUSSION EXERCISE 12.1

1. What is the relationship between the two clauses in each of the following sentences?

    Show up for work tomorrow or you're fired.

    Potatoes are good with chicken and rice is good with lamb.

    He's very strong, but he can't lift that boulder.

    The child ran into the room and she began to cry.

    They refuse to leave, for they love their country.

    The economy is getting better, yet the work force is shrinking.

    Study hard and you'll pass the test.

    You cannot take piano lessons nor can you take up the saxophone.

2. What is the connection between the two clauses joined by the correlatives in the following sentences?

Either we must contact the police or we must take the law into our own hands.

Neither did he return my calls, nor did he acknowledge my letter.

It is important to recognize that regardless of which coordinating conjunction is used, or the particular relationship that is expressed between the clauses, the clauses themselves have equal grammatical status and the conjunctions may be thought of as the glue that holds them together. What this means will become clearer to you in the next section, as we look at clause combinations in which the clauses do not have equal grammatical status.

SENTENCE BUILDING THROUGH SUBORDINATION

The process of **subordination** allows one clause to become a grammatical part of another. In this case the relationship is unequal, as the name of the process suggests. Consider the following clauses, for example:

(9)   The clown arrived. The children squealed with joy.

A much more natural way to express these two thoughts is as the one sentence of (10):

(10)   When the clown arrived, the children squealed with joy.

In this case, we have incorporated the first clause of (9) into the second. The second clause is the main idea that we are expressing, while the first is used to tell the time of the action of the second. In other words, it functions in much the same way that an adverb of time might. In this sense we say that it has become a grammatical part of the second clause.

English has a number of different ways in which one clause might be incorporated into another; the rest of this chapter will explore some of them in detail, showing the different grammatical roles that clauses can play and the unequal relationships that can hold among clauses of the same sentence. It will be helpful to have some terminology with which to refer to the clauses within a sentence. Unfortunately for us, many different terms for the same thing are current in grammatical description. For example, the clause that expresses the main idea is known as the **main clause** or the **matrix clause** or the **superordinate clause** or the **independent clause**. Any clause that performs a grammatical function within the main clause is known as a **subordinate clause** or an **embedded sentence** or a **dependent clause**. In this chapter, we will refer to them as *main clauses* and *subordinate* clauses. In our examples, we will sometimes use brackets to mark off the boundaries of clauses within a sentence and to show how clauses can nest inside other clauses, as in (11):

(11)   [[When the clown arrived] the children squealed with joy]

When we are analyzing the clauses of a sentence, it is important to keep in mind that subordinate clauses don't necessarily look exactly the way they would look if they were main clauses. They are clauses by definition as long as they have their own subjects and predicates, but they often undergo some distortion when they are subordinate. An introductory word may be added, or a word may be replaced or the verb may change its form. These distortions are very useful in helping listeners sort out the various clauses of the sentence, since they serve as red flags to alert listeners to the fact that the clause must be related grammatically to another clause that contains it. We will have more to say about these red flags as we discuss some common types of subordinate clause.

DISCUSSION EXERCISE 12.2

1. Place brackets around the subordinate clause in each of the following sentences:

   I realize that you have been waiting a long time.

   Rob bought the book after Connie had recommended it.

   He was not the same person to whom I had spoken earlier.

   That coffee contains caffeine is no secret.

   Although Sam wanted to leave, Ben convinced him to stay.

   Barry said that he would visit next week.

   We arranged for Mary to telephone her sister in Kansas.

   Before you fly off the handle, let me explain.

   For the men to leave without the women would be foolish.

   We discovered the baby whose parents had abandoned him.

2. Which do you think people understand faster, a. or b.? Why?

   a. They admitted you were right.

   b. They admitted that you were right.

3. Which of the underlined clauses are subordinate? What is your reasoning?

   <u>Theresa found the bone</u> that her dog had buried in the sand.

   <u>For Jake to announce his candidacy now</u> would be a mistake.

   <u>Jane finished the novel</u> although she was exhausted.

   After the first act, <u>Jim fell asleep</u>.

   I understand <u>that you can't come to my party</u>.

Each type of subordinate clause has its own identifying characteristics. In the next few sections, we will look at some of the most common types of subordinate clause.

ADVERBIAL CLAUSES

We began our discussion of subordinate clauses with the **adverbial clause**, the type of subordinate clause illustrated in (10). Any clause that behaves like an adverb falls under this category. Adverbial clauses all modify the verb of the main clause in some way: they may tell time, place, reason, or some condition placed on the action. There is always an introductory word, called a **subordinating conjunction**, that tells you the specific adverbial function of the clause. Some of these are shown in (12):

(12)   I'll see you [before the night is over].

She left [because she was angry].

He'll go [wherever life takes him].

We stayed [although we should have left].

Adverbial clauses are easy to spot. The subordinating conjunctions are reliable indicators of the function of the clause and, except for the introductory word, they look just like main clauses. They also move around relatively freely. All of the adverbial clauses in (12) could just as easily appear at the beginning of the sentence rather than the end. They might even interrupt the main clause, as in (13):

(13)   My sister, because she lacked confidence, failed at most things.

Adverbial clauses are often set off from the main clause by commas, particularly if they are long or if they appear at the beginning of the sentence or interrupt the main clause.

DISCUSSION EXERCISE 12.3

1. Identify the adverbial clause in each of the following sentences:

Since you have the time, you can proofread my paper.

We stopped to eat because we were hungry.

After the storm had passed, we surveyed the damage.

I haven't stopped thinking about you since you left.

We should look before we leap.

Although he is competent, he never finishes a job.

The men in my family, because they have big feet, have trouble finding shoes to fit.

While he was gone they hired someone to replace him.

All my children, since they are grown, support themselves now.

We waited until the parade passed through the town.

2. Subordinating conjunctions are sometimes made up of more than one word. Find the multiple-word subordinating conjunctions in the following sentences:

He will remain provided that we raise his salary.

I feel sorry for him even though he is mean.

Some people die so that others may live.

Can you think of any others?

3. What lexical category is *before* in the following sentence?

I'll meet you before the game.

Why isn't it a subordinating conjunction?

NOUN CLAUSES

There is another type of subordinate clause, one that behaves like a noun phrase. We call this type a **noun clause**. It will help to identify noun clauses if we refresh our memories about the functions that noun phrases perform in sentences. You will remember that most often they are either subjects of the clause, direct objects, indirect objects, objects of prepositions, or complements. Noun clauses perform some, but not all, of these functions. We find them occurring as subjects, direct objects, and complements. Let's look at the direct object function first. If we have a transitive verb like *know* in a sentence, it has to be followed by a noun phrase or a pronoun as its direct object. So all the sentences of (14) are possible, with *know* taking different kinds of direct objects:

(14)   I know the answer.
       I know the whole truth.
       I know my limits.
       I know Jim.
       I know them.

Now consider the sentences of (15):

(15)   I know [that you like me].
       I know [that the world is round].
       I know [that grammar can be difficult].

You'll notice that in each of these sentences the direct object is a whole clause, introduced by the word *that*. Because these subordinate clauses play the same role as a noun phrase in the sentences, they are called *noun clauses.* Because the specific role happens to be direct object, they are called **object noun clauses**. Noun clauses are often introduced by the word *that*, also considered a subordinating conjunction (but sometimes referred to as a *complementizer*).

DISCUSSION EXERCISE 12.4

1. Which of the following sentences have object noun clauses and which have adverbial clauses? How do you know? (Hint: a noun clause can be replaced with the word *something*.)

   We heard that you were leaving.

   She turned up the heat because she was cold.

   Let's buy food so that we can feed the baby.

   He suggested that I not ask questions.

   We regret that we cannot return the favor.

   I'll continue provided that you stop talking.

   Fay understands that the men will not help her.

   Joe anticipates that his team will win.

   Her husband resents that she earns more money.

   I've known her since she was a little girl.

2. Sometimes noun clause objects may omit their introductory *that*. Which of the following permit this omission? Did you find any disagreement?

   We know that air travel has become more dangerous.

   She regrets that she can't attend the meeting.

   I resent that he never speaks directly to me.

   We fear that the warring parties will never seek peace.

   He suspects that someone stole his keys.

3. Find one object noun clause and two adverbial clauses in the last paragraph before this exercise set.

4. Noun clauses may have special markings other than *that*. What marks the noun clauses in the following sentences?

   I wonder if they will arrive on time.

   Jan expected Steve to call her.

   Who knows whether the operation was a success?

   We arranged for Ted to receive the package.

   Lois doesn't know what will happen next.

   Can you think of others?

Now that you can identify object noun clauses, it should be easy for you identify **subject noun clauses**: they are noun clauses that play the role of subject in a sentence. We see in (16) a variety of different noun phrase subjects:

(16)   The answer is evident.

The truth seems obvious.

Your attitude was disturbing.

Mary's anger was apparent.

Again, we can replace any of these subject noun phrases with an entire clause:

(17)   [That he knows the answer] is evident.

[That she needs attention] seems obvious.

[That the project failed] was disturbing.

[That Mary was angry] was apparent.

In each case, the subject of the sentence is an entire clause, called a *subject noun clause.*

DISCUSSION EXERCISE 12.5

1. Identify the subject noun clause in each of the following sentences. Notice that the noun clause may be introduced by *that* or altered in some other way:

That they are angry is evident.

For you to leave now would be sensible.

Whether the plan will work remains to be seen.

What will happen next is anybody's guess.

That our resources are disappearing is sad.

2. Which of the following sentences have subject noun clauses and which have ad- verbial clauses? How do you know?

When you get there, give me a call.

If she finds my sweater, she can keep it.

That they have finally met is amazing.

For you to wait for me would be impractical.

Because he waited so long to seek advice, no one pitied him.

Whether or not you want me to be there, I will certainly show up.

Whether the tornado touched down is still unknown.

That Bud longs for the old days puzzles me.

For all of us to go would be silly.

Why you said that is a mystery.

3. Identify the subject and the object noun clauses in the following sentences:

That you criticized her indicates that you disapprove of her proposal.

That you didn't wait for me suggests that you didn't want me to come.

4. What is the structure of each of the following sentences? Use brackets to show the subordinate clauses.

That you answered before I could answer bothered me.

Tom knew that Jerry would come when he called.

You may be thinking that subject noun clauses sound very stilted and formal, maybe even awkward. You probably don't use them much in ordinary conversation, although you might use them more in writing. English provides us with another way to say these subject noun clauses so that they sound more conversational. We may move them to the end of the sentence, as illustrated in (18):

(18)   _____ is evident [that he knows the answer].

         _____ seems obvious [that she needs attention].

         _____ was disturbing [that the project failed].

         _____ was apparent [that Mary was angry].

Of course, when we do this, we leave the subject position vacant. English permits this only in the imperative; in all other situations, as we mentioned in Chapter 3, we must fill the subject position with a placeholder, or expletive. In this case we use the word *it*, giving us the sentences of (19):

(19)   It is evident that he knows the answer.

        It seems obvious that she needs attention.

        It was disturbing that the project failed.

        It was apparent that Mary was angry.

These are still subject noun clauses. They haven't changed their function, only their position in the sentence. They are now called **extraposed subject noun clauses**—literally, "put outside."

DISCUSSION EXERCISE 12.6

1. Identify the extraposed subject noun clauses in the following sentences:

It is imperative that you remain on board.

It would be futile for Holly to follow us.

It was important that he learn the truth.

It wasn't necessary for you to give me your seat.

It is evident that the tomato plants won't survive.

2. Put each of the extraposed clauses in the above sentences back in its original subject position.

3. Which of the following sentences contain object noun clauses and which contain extraposed subject noun clauses? How do you know?

John found that he couldn't talk to Alice anymore.

It was necessary for Jean to find work.

It was surprising that the stadium didn't fill for the home game.

My relatives learned that they couldn't visit without an invitation.

It startled her that the baby could talk at such an early age.

4. What is the structure of each of the following sentences? Place brackets around the subordinate clauses.

That you ate dinner before I arrived is obvious.

It is clear that he won't speak to me until I finish my homework.

5. Name the type of subordinate clause in each of the following sentences:

The teacher knew that the students were tired.

It was important for Linda to take that course.

After the crowd dispersed, the city workers swept the streets.

That snakes eat mice isn't surprising.

Let's party until the neighbors complain.

For Ollie to admit his guilt was a big step in his rehabilitation.

We expected that you would be surprised.

It didn't bother me that we had no money.

Chuck learned that his friends loved him.

Lulu kept smiling even though she was enraged.

There is one other common function of a noun clause. You can probably figure it out by comparing the sentences of (20) and (21):

(20)   My excuse is <u>insufficient funds</u>.

The reason she gave for resigning was <u>ill health</u>.

(21)   My answer is <u>that I have insufficient funds</u>.

The reason she gave for resigning is <u>that her health is failing</u>.

You will recognize the underlined portions of the sentences in (20) as subject complements, since they follow a linking verb and describe the subject. You can see that in (21) the underlined portions are also complements, playing exactly the same role in the sentence as the noun phrases in (20). But since they are complete clauses, we call them **complement noun clauses**.

DISCUSSION EXERCISE 12.7

1. Which of the following have complement noun clauses and which have object noun clauses? How do you know?

My answer is that I'm staying home.

Ann answered that she would not speak before a big crowd.

Joe's excuse is that he is shy.

Mike said that he would hook up my computer.

The fact is that I can't face him.

2. Name the type of noun clause in each of the following: subject, object, extraposed subject, or complement:

I accept that you won't ever marry me.

It is reasonable that we should save some money.

That diamonds appreciate in value is a fact.

The fact is that we are all dismayed.

Nicole knew that the jig was up.

RELATIVE CLAUSES

One other important kind of subordinate clause is the one that behaves like an adjective, in the sense that it helps describe or identify a noun phrase. We first encountered this clause type in Chapter 5, where we discussed pronouns. To review, consider the two sentences in (22):

(22) He is the architect. The architect designed the building.

English provides us with a way of incorporating the second into the first by creating a **relative clause**, which describes a noun phrase in the main clause. This noun phrase is known as the head of the relative clause. Relative clauses, you will remember, use relative pronouns that replace the repeated noun phrase. The choice of pronoun depends on whether the noun phrase it replaces is human or not, and if it is human, whether it is a subject, an object, or a possessive within the relative clause. By those criteria, each of the following will require a different relative pronoun:

(23) He is the architect.   The architect designed the building. (who)

The firm hired the architect. (whom)

The architect's brother was hired by the firm. (whose)

This is the building.   The architect designed the building. (which)

You will also remember that the relative pronoun *that* can be used in place of *who*, *whom*, and *which*. In fact, some style manuals require the relative pronoun *that* instead of *which* under certain conditions. We will say more about the use of *that* and *which* in the next section. Finally, when we form relative clauses, we must put the relative pronoun at the beginning of the clause.

DISCUSSION EXERCISE 12.8

1. Create sentences with relative clauses by incorporating the second of these clauses into the first:

   She forgave the women. The women insulted her.

   My uncle is the person. You met the person.

   The child is absent. The child's seat is empty.

2. What is the head of the relative clause in each of the following?

   The person who just left is a foreign dignitary.

   Fran identified the thief who stole the computer.

   The lawyer whom you consulted is the best in town.

   Children who have bad colds should not be in school.

   Carl dreamed of a woman who would accept him for himself.

3. Complete the following instructions for making relative clauses:

   1. Replace the repeated noun phrase with _____

   2. If the replaced noun phrase is a human subject, use the relative pronoun ____

   3. If it is a human object, use the relative pronoun _____

   4. If it is a human possessive, use the relative pronoun _____

   5. If it is nonhuman, use the relative pronoun _____

   6. You can use the relative pronoun *that* in place of _____

   7. Make sure the relative pronoun is positioned _____

4. How would you need to revise the above rules to accommodate the following?

   He is the man to whom you were speaking.

   This is the book for which I was waiting.

   This is the scientist with whom she collaborated.

5. What are some less formal ways of expressing the relative clauses illustrated in 4.?

Although most of our examples so far suggest that relative clauses typically occur at the end of the sentence, they can occur following any noun phrase in the sentence and may interrupt the main clause to do so, as illustrated below:

(24)   The person [who left this note] has a sense of humor.

   The animals [that suffered the most] were the bears.

   The artist [to whom you are referring] lives in New Mexico.

DISCUSSION EXERCISE 12.9

1. Make sentences with relative clauses from the following pairs of clauses:

   I met the man. The man won the lottery.

   The girls are sensible. The girls run the store.

   You lent me a book. The book was interesting.

   The woman was proud. The woman's child won the spelling bee.

   We consulted the engineer. You recommended the engineer.

2. Which are relative clauses and which are noun clauses in the following? How do you know?

   We know that you are sad.

   He recognized the man that you described.

   They found the child that ran away.

   I find that too much work makes me irritable.

   Paul used the coffeemaker that makes ten cups.

RESTRICTIVE AND NONRESTRICTIVE RELATIVE CLAUSES

One thing we have not acknowledged yet is that there are actually two different kinds of relative clauses. All the examples we have given so far are of one type: they help to identify the head noun phrase. Without the relative clause, our listeners would not have enough information to know what the head noun phrase referred to. These relative clauses are called **restrictive relative clauses**, because their job is to restrict the head noun phrase enough so that you can identify it. But now consider the sentences of (25):

(25)   We visited Greece, which is a lovely country.

   My friend Mary, whose brother competed in the Olympics, is a good athlete.

   That doctor, whom I called repeatedly, refused to renew my prescription.

In these sentences, the head noun phrase has already been fully identified. The relative clause gives more information about it, but that information is not required for me to know what you're talking about. For example, in the first sentence, I would know which Greece you are talking about even without the relative clause. You are just telling me, additionally, that it is a lovely country. These relative clauses are called **nonrestrictive relative clauses**. As you can see, nonrestrictive clauses are set off from the rest of the sentence by commas, while restrictive clauses may not be set off by commas. If we take out a nonrestrictive relative clause, the basic meaning of the sentence remains intact, but if we remove a restrictive relative clause, the meaning changes. You can see this by comparing the two sentences of (26) and imagining what they would be like without their relative clauses.

(26)   The person who played Gandhi in that movie is a wonderful actor.

Ben Kingsley, who played Gandhi in that movie, is a wonderful actor.

There are also some differences in relative pronoun use between restrictive and nonrestrictive relative clauses. The relative pronoun *that* can only be used with restrictive clauses. So we could say the first sentence of (27), but not the second:

(27)   The person that played Gandhi in that movie is a wonderful actor.

*Ben Kingsley, that played Gandhi in that movie, is a wonderful actor.

Also, as noted earlier, some style manuals do not permit *that* and *which* to occur interchangeably; *that* is required in restrictive clauses, and *which* is required in nonrestrictive clauses. For example, *The book that you recommended is interesting,* but *A Tale of Two Cities, which you recommended, is interesting.*

It is very important to recognize that most of the time whether a clause is restrictive or not is not a matter of grammar; it depends on what you mean to say and what you think your listener already knows. For example, look at the pair of sentences in (28).

(28)   The Americans who have a lot of money travel extensively.

The Americans, who have a lot of money, travel extensively.

In the first sentence, I am talking only about a certain subset of Americans. But, in the second, I am talking about all Americans. The relative clause is the same, but I am using it for two different purposes.

DISCUSSION EXERCISE 12.10

1. Consider the following sentence:

Students who are intellectually curious love this course.

What would be the change in meaning if the relative clause were nonrestrictive?

2. Is this relative clause more likely to be restrictive or nonrestrictive?

The Joe Louis Arena [which is in Detroit] hosts many sporting events.

Under what circumstances could it be interpreted as the other one?

3. Identify all the relative clauses in the following sentences. Set off the ones most naturally interpreted as nonrestrictive with commas.

My sister Mary who lives in Florida is coming to visit.

The girl who had the most points won the contest.

Vancouver which is in British Columbia is a lovely city.

Ulysses S. Grant whom we studied in history class was a general.

All the people to whom you gave tickets can now see the show.

The thing that bothers me most is the humidity.

The child whose mother was late was getting nervous.

My parents who own a hardware store know how to fertilize lawns.

His new computer which he bought last week doesn't work.

The books that are on the shelf are for you.

4. The requirement of some style manuals that prohibits *which* from occurring in restrictive relative clauses has limited application. To see what the limitation is, try to construct restrictive relative clauses from the following without leaving a deferred preposition.

The meeting was canceled. You sent me to the meeting.

The dictionary was flawed. He worked with the dictionary.

REDUCED RELATIVE CLAUSES

English also contains structures known as **reduced relative clauses**. Relative clauses may be reduced, or shortened, in two different ways. The first is that the relative pronoun may be omitted, provided that it is an object within the relative clause. In the sentences of (29), for example, the relative pronoun is optional.

(29) I contacted the lawyer (whom) you recommended.
      He cooked the meat (that) we brought.

In each of the above sentences, the relative pronoun stands for an object in the relative clause and can therefore be omitted. Subject relative pronouns cannot be omitted by most speakers of English, as you can see in (30):

(30) I contacted the lawyer who handled your case.
      *I contacted the lawyer handled your case.

      He cooked the meat that was in the refrigerator.
      *He cooked the meat was in the refrigerator.

DISCUSSION EXERCISE 12.11

1. Which of the following relative clauses can be reduced?

   Here is the book which I read last night.

   The man whom she married is a good husband.

   The person who delivers our mail is late again.

   The doctor whose patient survived the operation was elated.

   The plan that you proposed was accepted by the board.

2. Does this reduction apply equally to restrictive and nonrestrictive clauses? Test it on the following sentences:

   Mary, whom you met in Kentucky, is my cousin.

   The Eiffel Tower, which we visited on our European tour, is magnificent.

The other way in which a relative clause can be reduced is by the elimination of the relative pronoun and a following form of the verb *be*. The pairs of sentences in (31) show how this reduction works.

(31)   I like the person <u>who is</u> sitting next to you.

      I like the person sitting next to you.

      A man <u>who was</u> leaving the building noticed the fire.

      A man leaving the building noticed the fire.

Because the words eliminated from the clause are often *who is*, this reduction is sometimes called **whiz-deletion**. But the reduction can apply to any relative pronoun followed by any form of *be*, as illustrated in (32):

(32)   Omit the words (which are) underlined in red.

      They finally bought the house (that is) on the hill.

DISCUSSION EXERCISE 12.12

1. Does whiz-deletion work for nonrestrictive clauses as well as restrictive? Use the following as a test:

   Montreal, which is a large and beautiful city, is located in Quebec.

   My friend Alice, who is the first in her family to attend college, is very smart.

2. Which of the following relative clauses can be reduced in standard English?

   The woman who is rearranging your furniture looks worried.

   I studied the documents that you sent to me.

   My friend Max, whom you met last summer, just won the Nobel prize.

   This is the child who has chicken pox.

   I translated the letter that was originally written in German.

3. For the clauses in Exercise 12.12.2. that cannot be reduced, explain why not.
4. Can you find an example of whiz-deletion in the paragraph preceding this exercise set?
5. What additional step must be taken if all that is left after reduction is an adjective, as in the following?

    The horse that is brown won the race.

You'll notice that reduction of relative clauses gives us two ways to describe some sentences. If you describe a sentence such as (33) as a reduced relative clause,

(33)   They visited the ruins in the ancient city.

then you are implying some comparison to another sentence in which the clause is not reduced: *They visited the ruins that are in the ancient city.* You are also making reference to some mental process that leads from one to the other. But it is also possible to describe the end result of this process. In that case, we would say that *in the ancient city* is a prepositional phrase that serves to modify the noun phrase *the ruins*. We saw the same possibilites for different kinds of description when we talked about ellipsis in clause coordination in Chapter 8. When something is omitted that listeners can restore in order to understand the sentence, we can describe the relationship between the restored and unrestored version, and we can also describe the structure of what people actually say. Using both kinds of description gives us a fuller, richer picture of how the language works.

DISCUSSION EXERCISE 12.13

1. Describe each of the following sentences in relation to another, more complete sentence to which it is related:

    The person sitting next to me is breathing heavily.

    The house on the corner is vacant.

    A woman rejected by that employer filed a discrimination suit.

    The baby, startled by the noise, began to cry.

    The startled baby began to cry.

2. Describe each of the above sentences in terms of its structure as it is said, without reference to mental processes.

Combining clauses by coordination and subordination gives us infinite variety in the kinds of sentences we can construct. Even everyday conversation exhibits an amazing amount of complexity and linguistic virtuosity on the part of speakers and listeners alike. The next chapter will explore in greater depth the sentence construction process.

REFLECTIONS

1. Demonstrate that *after* and *since* can be subordinating conjunctions or prepositions.

2. There is another kind of noun clause that we did not discuss in this chapter, sometimes called a *relative nominal*. Examples appear underlined below:

   <u>Whoever did this good deed</u> should be praised.

   I'll accept <u>whatever you offer me</u>.

   He'll go with <u>whoever invites him</u>.

   Why do you think the last sentence uses *whoever* rather than *whomever*, even though the preceding word is a preposition? To help your explanation, consider the sentence below, which does require *whomever*:

   He'll go with <u>whomever he invites</u>.

3. A common complaint among grammatical purists is the following sentence type:

   The reason I am tired is because I worked late last night.

   What do you think is the basis for the complaint? What would make this sentence strictly grammatical?

4. A common definition of an independent clause is "a clause that can stand alone." Can you show why this definition does not work well for sentences that contain noun clauses?

5. Can you explain the ambiguity of the following sentence?

   The philosophical Greeks valued learning.

6. The following sentence contains a relative clause reduced by whiz-deletion:

   The man tossed the ball tossed the ball.

   What makes it so odd?

7. In *if . . . then* sentences like the ones below,

   If you were in my shoes, then you would do the same thing.

   If she follows my instructions, then she will get there.

   Which clause is the subordinate one? Why? What kind of subordinate clause is it?

8. Because subject and object noun clauses behave just like noun phrases, they can be rearranged in the passive voice. What is the passive version of each of the following?

Everyone knows that Harry loves Sally.

That you acknowledge your mistakes impresses your colleagues.

9. A character in Toni Morrison's novel *Song of Solomon* (New York: A Signet Book, New American Library, 1977) says, "you're the only one knows so far," and "They're the ones brought me out here." In what way is his rule for reducing relative clauses different from the rule for standard English?

PRACTICE EXERCISES   (Answers on p. 275)   _____

1. Which of the following clauses are combined by coordination and which by subordination?

    1. Jerry lost his wallet, but he had some spare change in his pocket.

    2. Michelle bought things on sale, for she was a thrifty woman.

    3. We stayed overnight, since we missed the last train.

    4. Neither could she sleep nor could she eat.

    5. Judy worked hard although she expected no reward.

    6. Park your car somewhere else or you will get a ticket.

    7. Bill was happy because he won the prize.

    8. We missed the train, so we booked a hotel.

    9. The dogs sleep in the barn and the cats sleep in my bed.

    10. I'll sign the contract if you delete the first paragraph.

2. Tell whether the underlined word is a coordinating conjunction, a subordinating conjunction, a sentence adverb, or a preposition.

    1. Jill stopped <u>before</u> she got to the corner.

    2. My uncle played the guitar, <u>but</u> he hated the banjo.

    3. Read this book <u>and</u> you'll see what I mean.

    4. Politicians can't all be trusted; <u>however</u>, we must rely on them.

5. Let's have a drink <u>after</u> the meeting.

6. They danced <u>until</u> they fell to the ground.

7. She was very intelligent; <u>moreover</u>, she had good judgment.

8. The children were excited, <u>yet</u> they were reluctant to participate.

9. He's sad <u>because</u> his dog ran away.

10. I stayed up <u>until</u> dawn.

3. Identify the noun clause in each of the following sentences. Tell whether it is a subject noun clause, an object noun clause, or an extraposed subject noun clause.

1. It amuses me that you like doing housework.

2. For Karen to quit now would be foolish.

3. That she loves her children is apparent.

4. Betty knew that I was busy.

5. It would not make sense for your associate to redo your work.

6. My brother likes his friends to entertain him.

7. It seems clear that the plan is unworkable.

8. We regret that we can't see you more often.

9. The heat caused the animals to sleep a lot.

10. It is fitting that she receive the award.

4. Which of the following sentences contain noun clauses and which contain adverbial clauses?

1. The apartment is clean although it isn't elegant.

2. Liz accepted that she would never be a great opera singer.

3. For the crew to set sail tomorrow would be smart.

4. I'll participate provided that you listen to me.

5. The conductor knew that the train would be delayed.

6. We waited until the sun had set.

7. It would be wise for the children to remain in school.

8. Deb expected that the foundation would collapse.

9. The trash spilled because you forgot to seal the bag.

10. The interviewer will hire him if he makes a good impression.

5. Which sentences contain object noun clauses and which contain complement noun clauses?

1. Ted knew that Irene would pack a lunch.

2. The plan is that we will meet in New York.

3. My reason for being late is that my alarm didn't go off.

4. All teachers expect that students will learn.

5. His biggest fear is that no one will show up.

6. The president claimed that he would attend the reception.

7. Our expectation is that the paint will peel before long.

8. The sad part is that they used to be good friends.

9. Her sister found that the house had been destroyed.

10. The students decided that they would prefer a take-home exam.

6. Identify the relative clauses in the following sentences. What is the head of each relative clause?

1. I went to the branch library that is located in my neighborhood.

2. The person to whom you were speaking is a sailor.

3. The doctor whose patients complained was fired.

4. The painting that I mentioned is not for sale.

5. I need to consult with someone who understands my problem.

6. The baby that is crying needs to be fed.

7. One example that I gave you is wrong.

8. They identified the man whom she had accused.

9. This is the zoo which houses the pandas from China.

10. We met the couple who adopted the twins.

7. Identify the subordinate clause in each of the following sentences and tell what kind it is.

1. Since you're already here, I don't need to call you.

2. Chuck wanted Bill to play soccer.

3. It is imperative that you refrain from smoking.

4. The annoying thing is that I can't go home until spring.

5. That she admires you is obvious.

6. Carolyn said that she would support our decision.

7. Call me if you have the time.

8. The stranger who accosted you is now in jail.

9. That is the story that my grandparents told me.

10. For Chris to lie about her age would be silly.

8. Which of the following restrictive relative clauses could also be nonrestrictive? How would the meaning of the sentence change?

1. My cousin who lives in Denver is coming to visit me.

2. The boy that threw the stone apologized.

3. The young man who sat in the front row never smiled.

4. He spoke to his friend who raises sheep in Australia.

5. Toby fed the cat which he found in his barn.

9. Which of the following relative clauses can be reduced by whiz-deletion? Which can be reduced by omitting the object relative pronoun? Which cannot be reduced at all? Why?

1. Nancy, who is a very good dancer, just opened a studio.

2. She disapproved of the man whom her daughter married.

3. My aunt Charlotte, whom you've met, turns ninety this month.

4. Is he the actor who is starring in your new movie?

    5. Mr. Jenkins, who was seated next to me, began to snore.

    6. Every child that plays this game loves it.

    7. The trees that are marked with red dots will be cut down.

    8. The spider, which was caught in its own web, began to struggle.

    9. Can I borrow the necklace that you wore last night?

  10. This quilt, which she made by hand, is beautiful.

10. Reduce the following relative clauses and describe the structure that remains after reduction.

    1. The boy who was standing on the corner laughed at me.

    2. A woman who was insulted by the remark left the room.

    3. The shelters that were in the park were torn down.

    4. The nervous man, who was expecting a call, paced the room.

    5. The hat that is made of straw is my favorite.

13

SENTENCE ANALYSIS

In the last chapter, you were able to see how people use language creatively to make clauses nest inside other clauses to express complex ideas. Although we concentrated on one kind of subordinate clause at a time, you saw some examples of sentences in which more than one subordinate clause appeared, even one within another. In this chapter, we will explore a broader range of possibilities for putting sentences together and develop a method of describing the very complicated sentences that occur routinely even in ordinary conversation.

SIMPLE SENTENCES

Sentences that are made up of one clause are known as **simple sentences**. They are easy to analyze, of course, as long as you remember that *simple* is a technical definition that does not imply anything about the length of the sentence nor its complexity in other ways. For example, both sentences of (1) are simple sentences:

(1) Jodi laughed.

The tall dark-haired woman with the ribbon in her hair sought advice at the beginning of the semester from the students in my grammar class.

What makes a sentence simple is that it has one subject and one predicate—that is, a noun phrase and a verb phrase that form a constituent. The basic structure of both sentences in (1) can be described with a tree diagram such as the one in (2). For this diagram and those that follow, the abbreviations are

> S = sentence
> NP = noun phrase
> VP = verb phrase
> PP = prepositional phrase
> P = preposition

(2)

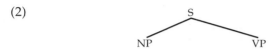

For the first, the NP and the VP are only one word apiece; for the second, the NP and the VP have multiple-word constituents within them. The fuller structure of the subject noun phrase in the second sentence can be represented by the diagram in (3).

(3)

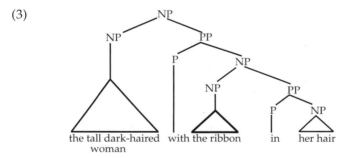

We have left out some of the details of the noun phrases, putting triangles in their place, but even as it is, (3) gives you a good idea of how structurally complex even one clause can be.

DISCUSSION EXERCISE 13.1

1. What is the structure of each of the following noun phrases? It might be helpful to put brackets around the constituents. Remember that constituents may nest inside other constituents.

   the house on the hill

   the building to the left of the church

   the answer to the first question of the exam on cell division

2. Draw a tree diagram of the verb phrase of each of the following sentences:

   I respect the students in my grammar class.

   We should meet at the end of the month.

   Look for it in the box under my bed.

3. Draw a tree diagram of the verb phrase in the second sentence of (1).

Tree diagrams such as the ones you drew for the exercises above are very useful tools for both describing and discovering the structure of sentences. Sentences themselves do not directly reveal their structure, since the words simply follow one after the other. But the tree structures show what groups into a constituent and how constituents nest inside other constituents. We know that everything in a sentence must be part of some constituent which, in turn, bears a particular relationship to the other constituents of the sentence. Creating a tree with all its interconnecting branches guides us in understanding how those constituents are related. The end result is a visual representation that lays out the multidimensional complexity of the sentence. You will better appreciate the value of the tree diagram as we begin to explore sentences that are not simple sentences.

COMPOUND SENTENCES

We have already discussed **compound sentences** at length. These are sentences made up of two or more clauses joined together by coordinating conjunctions. Each clause connected to another by means of a coordinating conjunction is considered to be a main clause and has equal status with the clause(s) with which it forms the compound. The sentences of (4) are some examples of compound sentences:

(4)   Spelling programs are useful, but they do not fix all spelling errors.

Neither can she sing nor can she play the piano.

His wife died years ago, yet he still dreams about her.

Although not always stylistically appealing, compound sentences may have more than two conjoined clauses, as in (5):

(5)   Mary laughed and Fred cried, but Conrad remained silent.
It rained on Saturday and it rained on Sunday, so we canceled the picnic.

DISCUSSION EXERCISE 13.2

1. Create compound sentences with the following coordinating conjunctions: *either . . . or*, *so*, *for*, and *or*.

2. You will remember that conjoined clauses can be reduced by ellipsis if they contain repeated elements. When they are so reduced, they may be described as *ellipted clauses*, or they may be described in terms of the remaining elements. Reduce the following conjoined clauses and describe the structure that remains:

Phil signed the letter and Marge signed the letter.

We identified the problem and we solved the problem.

The assistant collected the data and the assistant wrote the report.

3. Do the same descriptive options exist for these sentences once ellipsis occurs?

Dan wrote the music and Mike wrote the lyrics.

Mildred cooked the chicken and Jack cooked the fish.

Normally, sentences are only considered *compound* if enough structure remains for them to be recognized as having conjoined clauses. The sentences in Exercise 13.2.3. could be labeled *compound* after ellipsis, but the ones in Exercise 13.2.2. would not be.

4. Add a third conjoined clause to each of the following compound sentences:

Ken got there early but he couldn't find a parking space.

My friends came to my party and they cheered me up.

He didn't give money to charity nor did he volunteer his time.

Compound sentences can also be conveniently represented by tree structures, such as the one below, where cc is a coordinating conjunction:

(6)

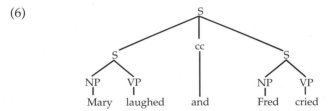

Sentences with more than two clauses, of course, will have more branches:

(7)

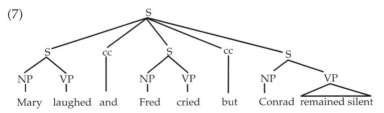

DISCUSSION EXERCISE 13.3

1. Draw tree diagrams for the sentences in (4). How did you decide to represent correlative coordinating conjunctions?

2. Draw a tree diagram for the second sentence of (5).

3. Give a sentence that fits the structure given below: (Remember that we use triangles instead of filling in the details of each clause.)

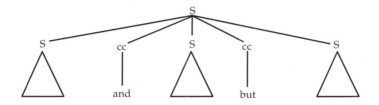

COMPLEX SENTENCES

Sentences that contain at least one subordinate clause (and no conjoined main clauses) are called **complex sentences**. These may contain any number of noun clauses, adverbial clauses, and relative clauses. We have seen many examples of sentences that contain one of these subordinate clauses. Below are some examples of sentences that contain two:

> (8) The teacher that won the award knows that her students appreciate her.
>
> Jean kept reading the book which John gave her until she fell asleep.
>
> A woman who happened to be there comforted the child who was upset.

Using the abbreviations NC (noun clause), AC (adverbial clause), and RC (relative clause), we can draw tree diagrams to represent the structure of these sentences. Let's begin with the subject noun phrase of the first sentence: *the teacher that won the award*. That is the whole subject noun phrase, but part of it is a relative clause describing the head noun phrase *the teacher*. Diagrammatically, it looks like this:

(9)

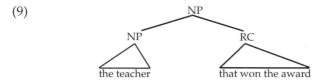

The verb phrase of the same sentence contains a transitive verb and an object noun clause, and can be represented diagrammatically as (10):

(10)

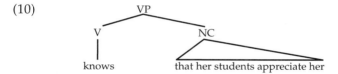

The clause structure of the whole sentence is shown in (11):

(11)

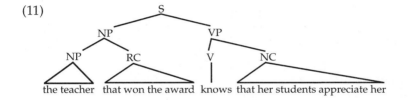

the teacher   that won the award   knows   that her students appreciate her

---

DISCUSSION EXERCISE 13.4

1. Give two different sentences with the following structure:

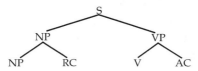

2. Give two different sentences with the following structure:

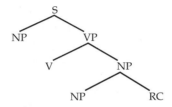

3. Draw tree structures for the remaining sentences of (8).

Sentences can have any number of subordinate clauses combined in a variety of ways. The sentences of (12) have three, four, and five subordinate clauses, respectively:

(12)   The student who was ill came to class because she knew that the assignment was due.

One person who was waiting noticed that the doctor that was on duty expected patients to pay their bills before she examined them.

It amazed Alice that the puppy which she just adopted had already learned that she would feed him before she prepared her own dinner if he wagged his tail.

It would not be easy for you at this point to describe the exact relationship that each of these clauses bears to the rest of the sentence, but native speakers of the language will understand the sentences without much difficulty and will be able to create equally complex sentences with similar ease. Much of human linguistic creativity involves the ability to embed many subordinate clauses into one sentence.

DISCUSSION EXERCISE 13.5

1. Identify the subordinate clauses in the sentences of (12).
2. Can you add another subordinate clause to the last sentence of (12)?
3. Which of the following sentences are compound and which are complex?

    I knew the answers but I froze on the exam.

    Everyone laughed when they came on stage.

    The boys who collected the most money were praised by the principal.

    The salad is wilted and the bread is cold.

    Although it was still early, Vickie felt that she should leave.

    That she became a doctor reflects well on her family.

    The nurse said that the patient who had malaria recovered.

    The concert was long over, yet the crowds remained.

    Either you should use the gift or you should return it.

    Until she pays her dues she should not expect that the union will support her.

If you try to analyze the structure of each sentence in (12), you will quickly realize that some of the subordinate clauses are nested within others. For example, in the first sentence, the verb phrase contains two subordinate clauses, one nesting within the other. The tree structure associated with this verb phrase is given in (13):

(13)

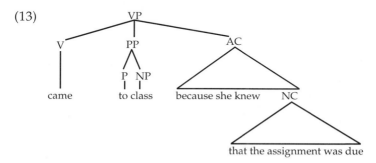

Each sentence of (14) also has one subordinate clause nested within another:

(14)   The cat that chased the rat that ate the cheese ran away.
            Before she learned that algebra was fun, Shelly hated math.
            I met the person who discovered that Einstein was a genius.

DISCUSSION EXERCISE 13.6

1. Identify the subordinate clauses of the sentences of (14) and tell which is nested within which. Use tree diagrams if they help you to see the structure.
2. Draw the tree structure of the subject noun phrase of the following sentence:
   Politicians who trust people who buy them presents are foolish.
3. Draw the tree structure of the following sentence:
   I believe that politicians who trust people who buy them presents are foolish.
4. Can you make the entire sentence above the object noun clause of a larger sentence? How many times are we permitted to repeat this process?
5. Can you insert another relative clause into the most deeply nested clause of the sentence in Exercise 13.6.3.? How many times are we permitted to repeat this process? What is the ultimate effect of multiple nestings?

Now consider the following sentences:

(15)   We knew that Len would succeed and Ben would fail.
   It surprised me that the gymnast could swim and the swimmer could jump.

The new element here is that the subordinate clauses are compound. Subordinate clauses, like any other constituent, may be conjoined with coordinating conjunctions to form compounds. The structure of the first sentence of (15) is as follows:

(16)

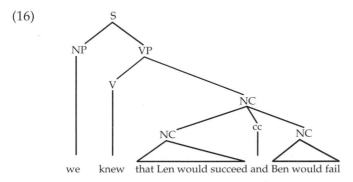

The sentence as a whole is still considered to be complex, since the compounding does not occur between two main clauses, but rather within a subordinate clause.

DISCUSSION EXERCISE 13.7

1. Which of the following sentences are compound and which are complex?
   The boys and the girls understood that they would not meet on Thursday.

> We suspected that the dog was scratching the wood and the cat was eating the plants.
>
> Either they will do the work or they will get fired.
>
> It is expected that no one will leave until the money is found.
>
> Lynn liked potato salad but Amy preferred cole slaw.

2. Draw tree structures for the sentences above, or place brackets around the clauses, and tell what kind of clause each one is.
3. Is the sentence in 13.7.2 compound or complex?

COMPOUND-COMPLEX SENTENCES

So far we have discussed simple sentences, compound sentences, and complex sentences. The one remaining type is a sentence which is compound while at least one of the clauses contains at least one subordinate clause. This type of sentence is **compound-complex**. The sentences below are compound-complex.

(17) Joel smiled but Phyllis knew that he was upset.

Mary was tired, so she left before the fireworks began.

We can represent these sentences diagrammatically as shown in (18) and (19), respectively:

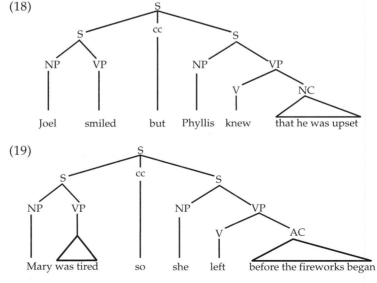

(18)

(19)

Every sentence in English is one of these four types: one clause alone or some combination of clauses, either bearing equal status to one another or one subordinated to another. Clauses can combine in infinite variety; they may get too complicated to be easily understood, but there is no fixed limit to the number of clauses that can be combined or the number of times clauses can nest within others in a particular sentence.

DISCUSSION EXERCISE 13.8

1. Which of these sentences are complex and which are compound-complex?

    We agreed that you would fix the faucet or that you would replace the sink.

    Karan knew the words to the song, but she forgot them when she started to sing.

    Tom did not file a tax return nor did he alert the IRS that he was leaving the country.

    The weather was rainy, yet we had a lovely vacation at the beach.

    The man who took the job never learned that he was considered temporary help.

2. Give a sentence which fits each of the following structures:

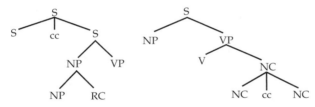

3. Tell whether each of the following sentences is simple, compound, complex, or compound-complex. Draw tree diagrams if they help you to visualize the structures.

    Sue found the jacket that she had lost before she moved to Florida.

    Neither will she accept responsibility nor will she recognize that we are displeased.

    The long, winding road to the top of the hill provides a beautiful view of the valley.

    Algebra is interesting but geometry is more challenging.

    Whenever she sees her mentor they have a discussion that lasts until midnight.

    Everyone loves clowns but mimes are not universally appealing.

    I left my job, for I always knew that I could do something more interesting.

    The reason that I'm late is that my alarm didn't go off, so I overslept.

    The weary traveler regaled us with tales of far-away lands and unfamiliar cultures.

    Expect the worst so that you will not be disappointed.

4. Draw tree structures for the first two sentences in Exercise 13.8.3.

We have now nearly completed our study of English grammar. At this point you should be able to describe any English sentence, from its highest level of organization to its lowest. You can determine how the clauses are arranged and what roles they play within the sentence. You can identify the constituents within each clause and describe the relationships they bear to one another, and you can identify the individual parts of speech that are the building blocks of sentences. You are also in a position to recognize many of the differences between standard English and its various nonstandard varieties and to understand the array of factors that lead people to use nonstandard grammar in their own usage.

This book has concentrated almost exclusively on formal written English and on that part of grammar that is concerned with how sentences are constructed, often referred to as the **syntax** of English. We will end our conversation about English grammar by talking about some of the features of oral English that have an impact on the way people ultimately construct their sentences and about the system we use to put English on paper.

REFLECTIONS

1. Choose one page of a textbook or a novel and label every sentence as simple, compound, complex, or compound-complex.

2. Find one piece of writing in which simple and compound sentences dominate the prose. Find another in which complex and compound-complex sentences dominate. What is your sense of the comparative difficulty of the two pieces?

3. One example sentence from this chapter is reproduced below:
   > It amazed Alice that the puppy which she just adopted had already learned that she would feed him before she prepared her own dinner if he wagged his tail.

   Read this sentence just once to someone and see if he or she (they!) can answer the following questions:
   > What amazed Alice?
   > Which puppy am I talking about?
   > What had the puppy learned?
   > Under what condition would Alice feed him?
   > When would she feed him?

   Would it surprise you to learn that people can understand sentences like these fairly quickly and easily?

4. How would you respond if someone asked you how many sentences there are in the English language?

5. Discussion Exercise 13.3 asks you to draw the structure of the sentence

    It rained on Saturday and it rained on Sunday, so we canceled the picnic.

    Here are two possible structures for this sentence. Which do you think better reflects the meaning of the sentence? Why?

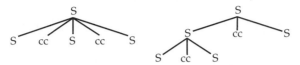

PRACTICE EXERCISES (Answers on p. 278) _____

1. Draw tree diagrams that represent the structures of these simple sentences. Use triangles instead of filling in all the details.

    1. My old friend from school won a million dollars in the lottery.

    2. The curve in the road caused an accident on that hill.

    3. The woman sitting on the sofa is president of the club.

2. Give a sentence that matches each of the structures below. What kind of sentences are these?

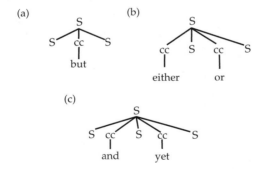

3. Identify the subordinate clauses in each of the following sentences. Tell whether they are adverbial, relative, or noun clauses. Which nest inside others? What kind of sentences are these?

    1. That you never smile suggests that you are sad.

    2. The sweater that the clerk showed me had a tear in it.

    3. People who are rich often do not appreciate the problems of those who are poor.

4. The boy whose mother died said that he wanted to live with his uncle.

5. The rules had been established before they arrived.

6. I consulted the lawyer whom you recommended because I knew that she would give me sound advice.

7. When you get to town, I suggest that you visit the library that was just built.

8. The girl that laughed when you entered the room didn't realize that you were upset.

9. My relatives will shop wherever they can find the best bargains.

10. It is necessary for Bob to understand that we meant no harm.

4. Draw tree diagrams that correspond to the sentences above.

5. Give a sentence for each of the following structures:

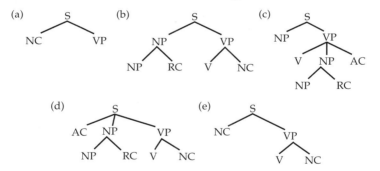

6. Give a sentence for each of the following structures:

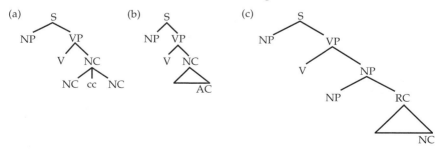

7. Which of the following sentences are compound and which are complex?

  1. The gift that you bought me for my birthday is beautiful.

  2. She knows that you left because you were tired.

  3. We sometimes like to ski in winter, but we prefer to keep warm.

  4. Either he will call or he will write.

  5. That you drive so fast worries me, because I love you.

  6. Judy expected Tom to show up for rehearsal.

  7. Pat thinks that grammar is easy and algebra is difficult.

  8. Call your mother and tell her the news.

  9. It is unlikely that you will get special treatment.

  10. You may not have a cookie nor may you eat potato chips.

8. Give a sentence that matches each of the structures below. What kind of sentences are these?

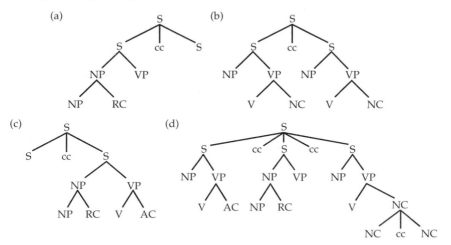

9. Draw a tree diagram for the following sentence:

  The woman who wrote the book met with the man who did the illustrations and they decided that they would do nothing until they heard from the publisher.

10. Label each of the following sentences as simple, compound, complex, or compound-complex.

   1. Everyone in my family appreciated your kind hospitality last night.

   2. It bothers us that we can't earn more money at this job.

   3. Although it rained, we all had a good time at the picnic you planned.

   4. My cousin Sue, who is also my best friend, stayed until everyone left.

   5. The cup broke when the cat landed on the table, but I was able to fix it.

   6. His excessive frankness will get him in trouble with his co-workers.

   7. Lend me fifty dollars and I'll never ask for another loan.

   8. Her instincts told her that she should get up and leave.

   9. I need a rest, so I will spend a week in the mountains.

   10. Since you're free and since you'll be around anyway, I'm inviting you to my party.

14

WORD CONSTRUCTION, PRONUNCIATION, AND SPELLING

Although the main focus of this book has been on the way in which English organizes words into sentences, or English syntax, we must remember that syntax does not exist in a vacuum. As speakers of English, we know much more about our language than how to put words together to make sentences. Other important dimensions of our knowledge are word construction and pronunciation. In addition, most of us also know how the language is written. Although it is beyond the scope of this book to explore word construction, pronunciation, and spelling in depth, there is good reason to have some discussion about all of them. Since people's behavior in these areas may vary, many of the central concerns of this book surface in these areas as well: Which is standard? Which is correct? Who decides? On what grounds? How do these judgments change over time? When people ask questions about English and English usage, their curiosity naturally extends to matters other than syntax. And, as is the case for syntax, we find that there are no simple answers to these questions, which make them interesting for discussion and for learning about language as a feature of human behavior.

HOW ARE WORDS CONSTRUCTED?

We began our discussion of word construction in Chapter 2. You will recall that English words are composed of roots, either alone, or with affixes attached to them. Some of those affixes, called derivational affixes, may change the lexical category of a root. Others, called inflectional affixes, add grammatical information to the root and do not change its lexical category.

So, for example, we may think of the word *national* as a combination of the root *nation* (a noun) and the derivational suffix *-al*, which turns it into an adjective. To *national* we can add another derivational affix, *-ize*, which will turn it into the verb *nationalize*. If we then add the inflectional suffix *-ed*, we will add the grammatical information "past tense," but we have not changed the lexical category, since *nationalized* is still a verb. The study of how words are constructed is called **morphology**, and the individual pieces of meaning or grammatical function that make up words are called **morphemes**. Those that may stand alone as words, like *nation*, are called **free morphemes**; those that must be attached to another morpheme, like *-ize*, are called **bound morphemes**.

DISCUSSION EXERCISE 14.1

1. Divide the following words up into their component morphemes: *solidify, merriment, international, overbearing, magnify, indivisible.*

2. Which of the above morphemes are free and which are bound? Are roots always free morphemes? Are affixes always bound morphemes?

3. For each morpheme you identified in 14.1.1., give another word in which it occurs.

4. Some words have two roots: *firefly, blackboard, butterball.* What is the traditional term for such words?

We may think of speakers of English as having a mental dictionary, or lexicon, consisting of all the morphemes of the language. When we want to express meaning, we draw on this stock of morphemes, putting together the pieces of meaning that add up to the meaning we want to express. Similarly, when we see or hear a word, we can figure out its meaning by figuring out the meaning of each of its morphemes. We can also draw on our stock of existing morphemes to create new words. Although this is not the only way new words come into English, it is a very common way of enriching the vocabulary. If we know, for example, that the morpheme *tele-* means "long-distance," and *phone* means "sound," we can name the invention that transmits sound over a long distance a *telephone*.

DISCUSSION EXERCISE 14.2

1. What are some other technological developments that use the morpheme *tele-* in their names? What are the meanings of the morphemes it combines with?

2. People vary considerably in the size and nature of their mental lexicons. Speakers of English share a common core of morphemes, which enables them to communicate with one another, but their vocabularies will differ according to their individual experiences: their professions, their levels of education, or their hobbies, for example. Can you think of any specialized vocabulary that you know that is not likely to be shared by others in the class?

3. Many of our scientific and medical morphemes are borrowed from Latin or Greek. Can you divide the following words into their component morphemes and tell what each morpheme means? *zoology, rhinoplasty, pediatric, gynecology, orthodontia.*
4. Why is it difficult to sort out the component morphemes of *smog* and *brunch*?

You will have the impression from what we have said so far that making words in English is simply a matter of selecting the appropriate morphemes and lining them up in the right order. Although that is often the case, it can be more complicated because not all sounds are allowed to occur next to each other. If the end of one morpheme and the beginning of the next is not a permissible sound sequence in the language, we cannot put them together. In other words, our word-construction processes (*morphology*) may come into conflict with the ways sounds are allowed to pattern in our language (**phonology**). For example, as was noted in Chapter 11, we have a bound prefix *in-* that means "not." You will see in the words of (1) that it combines readily with root morphemes that begin with vowels:

(1)   inadvisable, inoperable, inaccessible, inedible, inexplicable

It also combines with root morphemes that begin with certain consonants, as illustrated in (2):

(2)   intolerant, indecent, insecure, inconsequential

But it ordinarily will not combine with a root that begins with *b, p,* or *m*. Instead, we use the prefix *im-*, as illustrated in (3):

(3)   imbalance, impolite, immature

We have not used a different prefix morpheme in (3); rather, we have shaped *in-* so that it fits better with the following sound. If you say the words *balance, polite,* and *mature* aloud, you will see that you must use your two lips to form the initial sound. The *n* of *in-* molds itself so that it also uses two lips. The result is *m*. Since *in-* and *im-* have the same meaning and function, we call them **allomorphs** of the same morpheme, which are variations of the same morpheme.

DISCUSSION EXERCISE 14.3

1. What are some other root morphemes that require *im-* rather than *in-* as the negative prefix?
2. The prefix *con-*, which means "with," occurs in the words *confer, consider, contemplate, conclude, concur, conjoin* (you will not necessarily recognize the roots of these words). Can you think of any words that require *com-* instead? Why?
3. We used the term *phonology* above. What are the meanings of its component morphemes?

The allomorphic variation that we see illustrated above for *in-* and *com-* is typical of the kinds of adjustments that are made routinely in English words. Understanding the restrictions on what sounds are allowed to occur together is essential to understanding how people create words from their stock of morphemes. In order for us to better understand the word-construction processes at work in our language, we need to say some more about English phonology.

WHAT IS PHONOLOGY?

When we talk about the *phonology* of any language, we are referring to all those elements of the language that relate to the organization of sound. Each language uses a subset of all the possible human speech sounds and arranges them in certain ways, allowing certain sequences and disallowing others. Some of those restrictions are obvious and show up clearly in our spelling system, as was illustrated above. Others are more subtle and harder to notice. For example, the sound *k* in English actually has more than one pronunciation, although native speakers of English do not pay conscious attention to these variations. Consider the pronunciation of the *k* sound at the beginning of the words of (4) (it is often spelled with the letter *c*):

(4) cook
     car
     koala
     caution
     crane
     cool

Now compare these words to the words in (5):

(5) kill
     cat
     keen
     cake
     kept

The initial sound in the words of (5) is a slightly different sound. Your tongue is in a different position and your lips have a different shape. You will notice the difference if you get your mouth all ready to say *cook* and then decide to switch to *kill* at the last minute. If you don't make some adjustment in the position of your tongue and lips, the word *kill* will sound rather strange, although it will still be recognizable as *kill*. Which *k* sound you use depends on the quality of the following vowel, so again this is essentially a restriction on what sounds are permitted to occur in sequence.

DISCUSSION EXERCISE 14.4

1. Which of the following words have a *k* sound like those of (4) and which have a *k* sound like those of (5)? *coat, cost, kitchen, kettle, cup*

2. Here is another example of variations of a sound that we tend not to notice as native speakers of English. Say the following words: *pot, pear, report, appear*. Pay special attention to the sound of *p*. Now say the following words: *happen, apple, spill, spot*, again paying attention to the *p* sound. Do you notice a difference in the pronunciation in the first group and the second? What is the difference? If they sound the same to you, try saying them with a piece of paper in front of your mouth. Does that reveal a difference to you?

Speakers of a language know which sounds are variations of the same basic idea of a sound. The idea of a sound is called a **phoneme**; the variations of that sound are called its **allophones**. *P* is a phoneme of English with at least two variations, or allophones, as in *pear* versus *apple*. *K* is also a phoneme of English with at least two variations, as in *kill* versus *coal*. If someone puts these sounds in the wrong place in a word, we will perceive their (his or her!) pronunciation as foreign or non-English, but these variations cannot change the meaning of the word. In other cases, however, we recognize sounds as meaning-bearing, and using them will change the meaning. So, I will not change the meaning of *kill* no matter which allophone of *k* I use, but if I use *g* instead, and say *gill*, it becomes a different word. *G* in English is a separate phoneme from *k*. We also say that those two sounds are in contrast.

Some phonological rules of a language tell how the phonemes may line up in sequence. That will differ quite a bit from one language to the next. Spanish, for example, does not permit a word to begin with an *s* if the next sound is a consonant. English, of course, has no such restriction, so we may say *score, state*, and *spill*, for example. English does not allow a word to begin with *m* followed by *b*, but Swahili does. English does not permit a word to end with the sequence *pf*, while German does. Some languages generally do not have much tolerance for consonants clustering together, especially at the ends of words, and they will interrupt their consonants with vowels. Other languages, like most dialects of English, have a high tolerance for consonant clusters. Consider the pronunciation of *twelfths*, for example, or *sixths*!

Other phonological rules of a language tell where to place the allophones of a phoneme in words. So, every speaker of English knows which particular variety of *k* goes in *kill* and which in *coal* or which version of *p* must be pronounced in *appear* and which in *spear*. There are many such rules in English, but as native speakers of the language, we apply them automatically and are not conscious of their existence. We do, however, recognize immediately if someone violates these rules, because it will sound odd or foreign to us.

DISCUSSION EXERCISE 14.5

1. Which of the following are permissible sequences at the ends of English words? (think of sounds, not spelling!) *mb, nd, lts, mp, mpst, kt, kd*

2. Which of the following are permissible sequences at the beginning of English words? *str, shtr, kn, tl, tr, bl, sb*

3. Japanese renders *The Wall Street Journal* as *woru-sutorito-janaru*. What do you think are some differences in the phoneme sequencing requirements of English and Japanese?

4. Consider the vowel sound of *bad, sad*, and *lad*. Compare it to the vowel sound of *band, sand*, and *land*. Do you detect a difference? What happens if you use one for the other? Are these different phonemes of English or allophones of the same phoneme?

WHAT IS PHONETICS?

Many (but not all) of the restrictions on sequencing are designed to achieve a certain compatibility between neighboring sounds. Compatibility is determined by aspects of sound production. The study of these aspects, or features, of pronunciation is called **phonetics**. The features of pronunciation are determined by the following:

1. the place the air exits
2. vocal cord activity
3. the nature of the obstruction of air
4. the organs that form the obstruction

Most speech sounds originate as air in the lungs; as the air escapes from the body it is obstructed in a number of ways, and that is what gives each sound its unique quality. Let's look at each of these features in turn.

The air used for making speech sounds can exit the body from two places, the mouth and the nose. If the air exits only the mouth, we call the sound an **oral** sound. If air is permitted to exit from the nose, we call the sound **nasal**. Try comparing the first sound of *bat* with the first sound of *mat*. You will notice that *b* is an oral sound, with no air exiting from the nose; *m*, on the other hand, is a nasal sound.

DISCUSSION EXERCISE 14.6

1. Which of the following words begin with oral sounds? Which begin with nasal sounds? *mill, bill, note, dote, coat*

2. Which of the following words end with oral sounds? Which end with nasal sounds? (Think sound, not spelling!) *lamb, song, lab, cotton, machine*

3. There are some purely nasal sounds in English (at least in American English) that carry meaning. See if you can get these meanings across by keeping your mouth closed and allowing air to escape only through your nose:

What did you say?

This is delicious!

Let me think about this.

Yes.

No.

I'm insulted.

I'm listening.

Another feature of speech sound is the presence or absence of vibration in the vocal cords. The vocal cords are muscles in your larynx. These muscles may be at rest or they may vibrate during the production of speech sounds. If they are at rest, we say the sound is **voiceless**. If they are vibrating, the sound is **voiced**. You can probably feel the vibration by holding your thumb and forefinger on your larynx. Say the sound *s*, which is voiceless. Then switch to *z* and you should feel the vibrations. You can also try saying the sounds with your hands over your ears, which will enable you to hear the vibrations in your head. All consonants can be made with or without vocal cord vibration. For example, for the *p* in *pet* the vocal cords are at rest. If you keep everything else the same but vibrate your vocal cords, the *p* will turn into a voiced *b*, as in *bet*. It will help to hear the difference if you try to say the two sounds in isolation, without the following vowel. The *p* will sound whispered, but the *b* will be accompanied by noise.

DISCUSSION EXERCISE 14.7

1. Which of the following words start with voiced sounds and which with voiceless sounds? *pat, bat, ten, den, kill, gill, sue, zoo, chill, Jill*

2. Which of these words end with voiced sounds and which with voiceless sounds? (There generally isn't as much vibration at the ends of words): *tap, tab, net, Ned, lick, big, itch, ledge*

3. What happens to each of the following words when you add vocal cord vibration to the final sound? *rich, bet, peck, bus*

A third characteristic of speech sounds is determined by how the air is obstructed. If you say the word *bet* again, you will notice that your two lips come together and obstruct the air flow completely. The air pressure builds up behind your lips, and when you open them, the air comes out in a burst. Any sound for which the air is totally obstructed is known as a **stop**. Now say the word *fish*, concentrating on the *f* sound. In this case the air flow is only partially obstructed. Your upper teeth and lower lip form a barrier, but the air continues to push through it in a steady stream. Sounds that are produced with such partial obstructions are called **fricatives**. Other examples of fricatives are *s*, *z*, and the two different *th* sounds of *thin* and *this*. A third type of sound is a combination of a stop and a fricative, called an **affricate**. Consider the first sound of *toy*: this is a stop. Now consider the first sound of *shoe* (*sh* is one sound): this is a fricative. Now say the *t* sound again, but instead of releasing air, release it as the *sh* sound. If you put the two sounds together as described, you get the first sound of *chair*, which is an affricate. The first and last sound of *judge* (again, think sound, not spelling) are also affricates.

DISCUSSION EXERCISE 14.8

1. Tell whether each of the following words begins with a stop, a fricative, or an affricate: *toy, chimp, silly, kiss, then, just, dog, get, van, thought, fill, zero, shoe, genre*

2. Tell whether each of the followng words ends with a stop, a fricative, or an affricate: *leave, myth, start, mass, kick, red, raise, bush, ridge, mirage, gruff*

3. Consider your pronunciation of the word *garage*. Does it end with a fricative or an affricate?

4. The *f* sound is a voiceless fricative. What sound does it become if you vibrate your vocal cords?

5. What is the difference between the first sound of *thin* and the first sound of *this*?

The last feature of sounds involves the organs that obstruct the air. The organs most often involved in the obstruction of air are the lips, the teeth, the tongue, and the roof of the mouth. For purposes of description, the roof of the mouth is divided into three areas: the alveolar ridge, right behind the upper teeth; the hard palate, the bony part of the top of your mouth; and the soft palate (or velum), the soft, fleshy part of your mouth, behind the hard palate. The point at which two organs come together to obstruct the air (either fully or partially or a combination of both) is called the **place of articulation**. Each place of articulation produces a different sound. So, for example, if you want to make a *t* sound, you put the tip of your tongue on your alveolar ridge. (Try it!) But if you raise the back of your tongue against your velum instead, you will make a *k* sound.

The sounds we have described so far are **consonants**, those sounds which have significant obstruction of air. All consonant sounds have four-part descriptions: they are either oral or nasal, voiced or voiceless, a stop, fricative or affricate, and they all have a particular place of articulation.

The first sound of *pipe* (*p*) and the first sound of *type* (*t*) can be described as follows:

	Place of Exit	Vocal Cord Activity	Nature of Obstruction	Place of Articulation
p	oral	voiceless	stop	two lips
t	oral	voiceless	stop	alveolar ridge

You can see that these two sounds are alike except for the place of obstruction. The sound *v*, as shown below, is different from both *p* and *t* in three of the four features of pronunciation.

	Place of Exit	Vocal Cord Activity	Nature of Obstruction	Place of Articulation
v	oral	voiced	fricative	upper teeth, lower lip

DISCUSSION EXERCISE 14.9

1. Which organs obstruct the air in the production of these sounds? *p, t, k, n, m, b, d, g, f, v*
2. Keep the place of articulation the same, but remove the vocal cord vibration from each of these sounds. What sound results? *b, d, g, z, v*
3. What consonant sound is described by each of the following?

    oral voiced stop, tongue on alveolar ridge

    nasal, lips obstruct air

    oral, voiceless, upper teeth and lower lip obstruct the air

    nasal, tongue on alveolar ridge

Of course, there are many sounds that are not consonants. **Vowels** are sounds that have virtually no obstruction of air, but rather shape the oral cavity with the lips and tongue to produce different vowel sounds. And, there is a class of sounds in between consonants and vowels, called **approximants**, that have some obstruction, but less than for the consonants. In English, the first sounds of *rid, left, yell,* and *wear* are considered to be approximants, phonetically speaking, even though we are used to thinking of them as consonants.

It is not necessary for you to know all the details of the production of speech sounds in order to understand that English, like all languages, puts restrictions on which sounds may occur next to, or sometimes just near, one another. These restrictions, in turn, affect the way we construct our words. We have already seen how the prefix *in-*, which ends in *n,* (made at the alveolar ridge) will not precede a root that starts with a sound made with the two lips. Instead, the *n* molds itself into an *m* for better compatibility between the sounds.

SOME RULES OF ENGLISH MORPHOLOGY

Let's return now to our discussion of word construction. We said earlier that to construct words we draw on our stock of morphemes to put individual pieces of meaning together. We also said that sometimes we run into a potential clash with what our phonology permits as a sequence. When that happens, morphemes mold themselves to their surroundings, giving rise to allomorphs of the same morpheme. The result is a compromise: we can still express the meanings we need to express and we keep neighboring sounds compatible as well: *in* + *tolerant* but *im* + *possible*. You will notice that what has happened is that the sounds next to each other have become more alike, in this case by a shift in the place of articulation of the first one. This process of making neighboring sounds more alike is called **assimilation**; it is probably the most common strategy we use to make neighboring sounds compatible. Let's consider another example of assimilation. Think about how we make nouns plural (only the regular ones). You might be inclined to say that we add an -*s* or an -*es*. This is true for spelling, but not for pronunciation. Say the words of (6) aloud, comparing the plural suffix of the (a) group with that of the (b) group.

(6) (a) cats
          tops
          socks
          cliffs

     (b) yards
          pubs
          rigs
          hives

You will hear that the first group adds an *s* sound, while the second group adds a *z* sound. Why would English go to the trouble of having two different pronunciations for its plural endings? Wouldn't just one be easier? The problem, of course, is that it is not easier in terms of pronunciation. If you consider the endings of the noun roots, you will see that the first group ends in a voiceless consonant, while the second ends in a voiced consonant. The choice of *s* (voiceless) or *z* (voiced) depends on the voicing of the final consonant. Again, we see assimilation at work. Two sounds next to each other share certain characteristics (in this case, the presence or absence of vocal cord vibration), in order to make them compatible in pronunciation.

DISCUSSION EXERCISE 14.10

1. Think about how we make nouns possessive in speaking. Remember: "add an apostrophe *s*" is a spelling rule. Use these possessives to help you focus on pronunciation: *Pat's, Jeff's, Rick's, Bud's, Harve's, Meg's, Bob's.*

2. What are the two allomorphs of the morpheme that means possessive? Why are there two?

3. What similarities are there between possessive and plural formation in English?

4. Think about the third-person-singular present tense of verbs. We have said in this book that standard English adds the suffix -s to the base form. Is this a statement about spelling or pronunciation? Use the following verbs to help you figure it out: *spits, laughs, lisps, picks; digs, grinds, leaves, curbs.*

As you can see from the examples above, assimilation plays a large role in the pronunciation of words, but that is not the only strategy at work to make neighboring sounds more compatible. Consider the plurals of the following nouns:

(7)   bench
      glass
      garage
      judge
      glaze
      bush

We have already indicated that to make the plural we have a choice of two allomorphs: s and z, depending on the ending of the noun root. Now we see in (7) that there is more to the formation of plural nouns. We will have trouble if we try to add s or z to these words, because the final sounds are too much like s and z and they will blend together. Assimilation would not help us here, because the sounds are already too much alike. What English does in this case is add a vowel between the ending and the suffix, thus breaking up the two consonants. This process of adding a sound to create an acceptable sequence is called **epenthesis**.

DISCUSSION EXERCISE 14.11

1. Do we ever use epenthesis in the possessive of nouns? Test it with the following possessives: *judge's, witch's, Butch's, Tess's, Roz's.*

2. Do we ever use epenthesis in the third person singular present tense of verbs? Test it with the following verbs: *catch, miss, buzz, lodge, rush.*

Another important way that English deals with a potentially unacceptable sequence of sounds is to drop one of the "offending" sounds. This process is called **deletion**. This is especially true when too many consonants cluster together for comfort. Although our spelling system holds onto these consonants tenaciously, in ordinary speech they often tend to disappear when they are in clusters. *Limb, bomb,* and *climb* are examples of final -*mb* clusters that were simplified a long time ago by deletion. Initial clusters were also simplified long ago: *loud* used to start with *hl-*; *knave* and *knight*, of course, started with *kn-*; *gnat* and *gnarl* started with *gn-*. As you would expect, the process continues into modern English. Although we may not be consciously aware of it, in casual conversation many of us do not pronounce all the consonants in words like *months, twelfths, sixths,* or *asked.*

DISCUSSION EXERCISE 14.12

1. What do these English spellings sugggest to you about older pronunciations of the language? *night, wrong, comb*

2. Why do you suppose that *skimmed milk* has come to be called *skim milk?*

3. What do you think triggers the common misspellings *use to* and *suppose to?*

There is another way that a language can adjust to a potential violation of its sound-sequencing rules: reverse the order of the two sounds that cannot occur in sequence. This strategy is called **metathesis**. In English there are nonstandard pronunciations like *perty* for *pretty, preform* for *perform, nuc-ye-lar* for *nuc-le-ar,* and *intregal* for *integral,* which reverse the order of sounds to create acceptable sequences for individual speakers. Although standard English does not show any direct evidence of metathesis, there are examples of words that have changed over time by reversing the order of sounds, like *ask. Aks* (often spelled *axe*) was once a standard pronunciation of this word.

VARIATIONS IN PRONUNCIATION

It should be clear from our discussion about word formation and pronunciation that there will be variation in people's usage. The rules for acceptable sequences of sounds are not absolute. They will change over time, they can vary from one dialect to another, and they can vary according to the formality of our speech. They can even vary from one individual to another. What speakers of all languages have in common are the same strategies for avoiding unacceptable sequences: we all have available to us assimilation, epenthesis, deletion, and metathesis to create acceptable sequences. (Also see Reflection 12 at the end of this chapter.) Not everyone applies these strategies in the same way; standard English accepts some results and not others and over time may change what it accepts. No one these days objects to *ask* over *axe,* while *nuc-ye-lar* and *perty* at this time are considered nonstandard by most of us.

DISCUSSION EXERCISE 14.13

1. Some people say *ath-e-lete* for *athlete.* What sequence of sounds do they find objectionable? What strategy are they using to avoid it?

2. Many children have problems with the word *spaghetti*. What strategy do they use to create a more acceptable sound sequence?

3. How do you pronounce the word *lantern*? Have you ever heard anyone say it with the *e* and the *r* reversed?

4. Have you ever heard someone pronounce the word *specific* the same as *pacific*? What pronunciation strategy is at work here?

5. In the eighteenth century, the author Jonathan Swift objected to the pronunciation of *disturbed* as *disturb'd* rather than *disturb-e-d*. Here is his reaction to the consonant cluster formed by leaving out the vowel: " a jarring sound, and so difficult to utter, that I have often wondered how it could ever obtain." [See Albert C. Baugh, *A History of the English Language*, 2nd ed., (New York: Appleton-Century-Crofts, 1957), p. 312.] What does Swift's comment tell us about changes in the acceptability of certain final consonant clusters in English over time?

We find some very interesting dialect differences that result from differences in acceptability of certain sound sequences. For example, there are some dialects of English that are called "*r*-less." You can hear this dialect feature on much of the east coast of the U.S. as well as in much of England. R-less dialects suppress the pronunciation of the *r* sound (deletion) unless it precedes a vowel. So, for a speaker of an *r*-less dialect, the *r* is not pronounced in words like *yard, barn,* and *farm.* It is also not pronounced at the ends of words like *mother, father,* and *butter,* unless a vowel follows. To speakers of *r*-ful dialects it may sound as though speakers of *r*-less dialects drop all their *r*s, but in fact the *r*s are dropped only to avoid certain sound sequences.

Another good example of differences that arise from sequencing restrictions is **African American Vernacular English (AAVE)**, a dialect spoken by many African Americans across the United States, usually in informal settings. This too is an *r*-less dialect, but even more interesting is its preference for avoiding consonant clusters at the ends of words, especially if those consonants are produced at the alveolar ridge: clusters including *n, t, d, s,* and *z,* are avoided when possible. We see here a situation in which the phonology of a dialect comes into direct conflict with the morphology of standard English. Consider the fact that much of our grammatical information is expressed as suffixes: noun plurals, noun possessives, third-person-singular present tense of verbs, and past tense of verbs. (See Reflections 8 and 9 at the end of this chapter.) Consider also that these suffixes are usually pronounced *s, z, t,* or *d.* Every time you add one of these suffixes to the end of a word that ends in a consonant, you create a consonant cluster. What's going to happen if the phonology of the dialect says "avoid consonant clusters at the ends of words"? AAVE has some interesting ways of avoiding the conflict without losing meaning.

In the case of the present tense of verbs, the suffix of the third-person-singular carries no independent meaning of its own; rather, it repeats information already available in the subject. In fact, you may recall from Chapter 4 that there was a time in English when all verb forms of the present tense had endings, which have gradually been dropped over the years. The one remaining is the third-person singular, still required by standard English, as in the sentences of (8):

(8)   The car runs well.
      Lani bakes her own bread.

Since the suffix is redundant, no meaning is lost if it is dropped. Thus, AAVE permits the dropping of the suffix to conform to its preference for avoiding consonant clusters at the ends of words. In AAVE, the base form is used for all persons and numbers of the present tense: *The car run well, Lani bake her own bread.*

DISCUSSION EXERCISE 14.14

1. AAVE eliminates the suffix on third-person-singular present tense verbs even when there is no potential consonant cluster, as in *he do, she see.* Why do you think this occurs?
2. Can you think of any old-fashioned English verbs that put endings on verbs other than the third-person-singular present tense?

The expression of the possessive is another way in which AAVE differs from standard English, again as a result of reconciling the requirements of morphology with the preferences of phonology. You will remember from our earlier discussion that adding the possessive suffix will create consonant clusters in the same way that adding the present tense verb suffix will: *man's, hat's, girl's.* For possessives, AAVE permits the dropping of the suffix as long as the noun phrases remain in the order *possessor + object possessed.* So, as shown in (9),

(9)   The man hat
      The hat is the man's.

the first may omit the suffix, whereas the second may not because the possessor follows the object possessed.

DISCUSSION EXERCISE 14.15

1. Which of the following would be an acceptable possessive in AAVE? *that girl book, my mother house, the house is my mother*
2. Change the following standard English sentences into AAVE:
      He admires my husband's car.
      The woman's cat kills mice.

There are several other situations in which AAVE avoids a consonant cluster at the ends of words where standard English is more likely to retain the cluster. For example, if some other word expresses plural meaning, plural suffixes may be dropped in AAVE: *five cent*, for example. If an adverb tells that the verb is past tense, the past-tense suffix may be omitted: *they talk yesterday*. Contracted verbs that carry no independent meaning may also may be omitted, so that standard English *that woman's angry* (from *that woman is angry*) can be said *that woman angry*.

AAVE is a very good example of how a preference for certain sound sequences can affect other parts of the grammar. In an effort to adhere to certain sound sequences (and avoid others), the dialect differs from standard English in some of its word order restrictions and its expression of tenses. It even allows certain sentences to occur without an overt verb. In no instance is there loss of meaning: either what is eliminated is redundant information or the meaning is expressed in some other way.

DISCUSSION EXERCISE 14.16

1. What would be the way to say these in AAVE?

    He needs two pencils.

    The girl's upset.

    The teacher's talking to Ed's father.

2. Dropping of endings in AAVE might extend to situations in which there would be no consonant cluster, such as *he angry* (for *he's angry*). Why do you think this occurs?

3. All speakers of English drop consonants in certain situations. Say these two sentences aloud at normal speed:

    I walked to school.

    I walk to school.

    Do they sound different? Say one of them to another person. Can they (he or she!) tell the difference?

STANDARD ENGLISH AND SPELLING

For most of us, spelling is an important dimension of standard English. Aside from some minor variations between American and British English, formal English does not readily accept variations in spelling, either dialectal or individual. We expect there to be one way, the right way, to spell words, and we expect dictionaries to tell us the right way; for many of us that is one of the major functions of a dictionary. We also tend to measure a person's level of education by his or her ability to spell words correctly. There is an agreed-upon standard in spelling to a much greater extent than there is for pronunciation or grammar. Such a standard did not always exist in English, as we mentioned in Chapter 1. It took several centuries to develop a universally accepted spelling system, but we are now at a point in our development where there is little controversy over the spelling of words.

This does not mean that spelling is easy for native speakers of English or for people learning English as a second language. One of our problems is that the oral language continues to evolve while the written language changes much more slowly. So some of our spelling problems stem from the fact that our writing system represents an older form of the language. For example, sounds that have dropped out of pronunciation may still have a written representation: *k̲nave, ni̲g̲h̲t, wif̲e̲, bom̲b̲*. In other cases sounds have changed their pronunciation but have a symbol that better represents the older pronunciation: *lau̲g̲h̲*, for example.

We must also keep in mind that many of the people who helped develop the modern English spelling system were scribes whose first language was French, and so we see the influence of French spelling in our alphabet: *g* with the sound of *j*, as in *gentle*; *c* with the sound of *s*, as in *citizen*, and many letter combinations for single sounds, such as *ch* in *chair*, *sh* in *shout*, and *th* as in *thin* and *this*.

The net effect of the evolution of the English writing system is that there is a very large gap between the spoken and the written language. Although we might have been told that written English is a direct representation of spoken English, we all know that "sounding it out" as a guide to spelling is often useless. English has "silent" letters, as we mentioned above. Also, the same symbol may be used for several different sounds (*c̲ake, f̲ather, m̲a̲n*), and different symbols can represent the same sound (*n̲e̲ed, l̲e̲af, machi̲n̲e*). These would be problems even if everyone spoke exactly the same way, but as we have seen, pronunciations can vary dramatically from dialect to dialect as well. Since there is only one agreed-upon spelling, the gap between the spoken and the written language is greater for some speakers of English than for others.

DISCUSSION EXERCISE 14.17

1. What spelling problems are represented in these words? *caught, laughed, colonel, diaphragm, receive, altar, alter*

2. Why do you think speakers of AAVE often omit the past tense *-ed* in writing?

3. Why do you think the words *merry, Mary,* and *marry* have three different spellings? (Students in different parts of the country will probably have different reactions to this question.)

SHOULD WE REFORM OUR SPELLING SYSTEM?

Because of the difficulty, and apparent irrationality, of English spelling, there have long been people who have agitated for spelling reform. George Bernard Shaw, for example, pointed out that *fish* could be spelled *ghoti: gh* as in *laugh, o* as in *women, ti* as in *nation.* Some people have argued that we need a phonetic spelling system, that words should be written as they are pronounced. When people say a phonetic system, they certainly don't mean truly phonetic, capturing every aspect of pronunciation. No one would argue that we should have two different symbols for the *k* sound in *cool* and the *k* sound in *kill;* nor would anyone want the *p* in *pill* to be different in writing from the *p* in *spill.* Allophones of the same phoneme are phonetically different, but we don't pay conscious attention to those differences. Trying to capture them in our writing system would be very difficult and would not serve any real purpose.

What spelling reform advocates really seem to be supporting is a *phonemic* writing system, one in which each symbol of the alphabet represents one phoneme of the language. By this we mean an alphabet in which the symbols represent the more abstract ideas of sounds rather than their actual pronunciation, because pronunciation tends to vary from one context to another. In such a system, each phoneme would have only one symbol and each symbol would represent only one phoneme.

DISCUSSION EXERCISE 14.18

1. Devise a phonemic spelling system for these words: *night, bite, rough, cuff, cake, break.*

2. Some people have likened changing to a new spelling system to changing to the metric system and argue that the practical disadvantages outweigh the advantages. What are the practical problems associated with such changes?

In many ways, the idea of a phonemic spelling system is appealing: people might be better spellers and English might be easier to learn to read (although these are empirical questions that would need to be tested). But from our previous discussions about variation, we know that English is not the same for all of its speakers. After whose phoneme system would the spelling system be modeled? For many people in the United States, the vowels of *cot* and *caught* are the same. Would we represent them as the same symbol? Would *pin* and *pen* be written with the same vowel or two different ones? Would *Mary, merry,* and *marry* have different vowels, or the same? Or, would two of them be the same and one different? Would the first vowel of *economics* be the same as the vowel of *beet* or the vowel of *bet*? Would *calm* be spelled with an *l* or not? Would *where* and *wear* have the same spelling? Whichever decisions we make, they will be better representations of some people's pronunciation than of others. We cannot have a purely phonemic alphabet for all of English because English itself is not completely uniform.

Another problem with the concept of a phonemic spelling system, of course, is that English will continue to change. Part of the reason for the irregularities in our current system is that pronunciation has changed, making some of the original choices seem irrational to us. For example, the letter combination *gh* originally represented a voiceless velar fricative, a sound like *k* but without total obstruction of the air flow. English lost this sound as a phoneme. Sometimes it stopped being pronounced (as in *night*); in other cases it changed to *f* (as in *laugh*). The final *b* in *bomb* was once pronounced, as was the initial *k* in *knight*. If we modernize our spelling system we will need to commit to periodic updates, for in time similar gaps between speaking and writing will resurface.

Those who object to changing our spelling system argue that a writing system does not need to represent pronunciation directly. Rather, it represents how we think about our language. Everyone accepts this principle at some level, since even a phonemic alphabet represents our ideas of sounds rather than how we pronounce them. But there are other facets of the mental organization of our language that are represented in our current spelling system. For example, many homonyms have different spellings that reveal their different meanings: *bare* and *bear*; *site, cite*, and *sight*; *meat, meet*, and *mete*. If all these words had the same written representation, it might make reading more difficult, not less. Furthermore, our current spelling shows that some morphemes with different pronunciations are related: *sign* and *signature, bomb* and *bombard, serene* and *serenity*. We would lose these visual connections if we wrote these words phonemically. For example, the first vowel of *sign* and *signature* would have different spellings.

DISCUSSION EXERCISE 14.19

1. Make up a phonemic spelling for *brood* and *brewed*. Do you think it would make written English easier if they were spelled alike? Why or why not?

2. The spellings *nite* and *lite* are often seen in advertising. Do you think they will become standard? Why or why not? Can you think of other common simplified spellings?

3. Why do you think there is resistance to spelling reform? What is your own position?

It has been the goal of this book to get you to think of English as a dynamic, fluid set of human behaviors. To the extent that it has succeeded in achieving its goal, you will not think of your study of English as over upon reaching this last page. Rather, you will be launched on a lifetime of listening to people using their grammatical knowledge. You will continue to keep your ear tuned to the subtle and not-so-subtle changes taking place in your own lifetime. You will relish the richness of the variety in English usage and you will appreciate the common goals and approaches to language that maintain English as a tool for communication for all of us: our preference for regular patterns, our strategies for reconciling the competing demands of phonology and morphology, and, above all, our fundamentally human need to speak our minds.

REFLECTIONS

1. Some morphemes may lose their meaning over time. When they do, they are no longer productive—that is, we cannot use them any more to make new words. One example of a nonproductive morpheme is *-ceive*, as in *receive, conceive, perceive, deceive.* Can you think of other nonproductive morphemes?

2. Words like *smog* and *brunch* are known as *blends.* Another such blend appeared in a local newspaper: *advertorial.* What do you think it is? Can you think of other recent blends?

3. What problem do the words *strawberry, gooseberry, cranberry,* and *huckleberry* present if you wish to divide them up into their component morphemes?

4. Sometimes people will create new morphemes by reanalyzing the morpheme structure of a word. For example, *hamburger* used to be an adjective that referred to the German city of Hamburg, and what we commonly think of as a hamburger was called *hamburger steak.* We have reanalyzed *hamburger* into *ham + burger.* What is the evidence for this? What is the meaning of *burger*?

5. Look up *garage* in a dictionary. What does it say about its pronunciation? Is one pronunciation favored over another in standard English?

6. In phonetics, the places of articulation are given technical names:
   bilabial: two lips
   alveolar: tongue on alveolar ridge
   labiodental: lips and teeth
   interdental: tongue between teeth
   palatal: tongue against hard palate
   velar: tongue against soft palate (velum)

   Thus, *p* would be an oral, voiceless, bilabial stop. What would *d* and *f* be?

7. In order to represent speech sounds, phonetics uses a special alphabet called the alphabet of the International Phonetic Association (IPA). If you want to learn this alphabet, consult any standard introduction to phonetics or linguistics.

8. The past tense in English exhibits assimilation. Think of the sound of the past tense suffix in *laughed, picked, ripped, missed,* and *talked.* Now compare its pronunciation to that of the suffix in *hugged, heaved, buzzed,* and *robbed.* How does this demonstrate assimilation?

9. The past tense exhibits another strategy for making acceptable sequences. Consider the past tense of these verbs: *list, raid, state, crowd*. What strategy is being used here?

10. In ordinary casual conversation, *have to* may sound like *hafta*. What makes the *v* of *have* become an *f*?

11. A clothing store in Michigan is displaying *"relax fit jeans."* By what strategy of pronunciation has *relax* become an adjective?

12. Although rare, another way English deals with neighboring sounds that are too much alike is to make them less alike. As you might guess, this process is called *dissimilation*. We see it in the suffix -*al*, a common derivational suffix that turns word roots into adjectives. But notice that in the words *angular, circular, cellular*, and *regular*, the suffix appears as -*ar*. What do you think is the explanation for this alternation between -*al* and -*ar*?

13. Loss of endings has characterized changes in English for many centuries, so it is not surprising that AAVE uses deletion as a strategy for resolving conflicts between phonology and morphology. It is also not surprising that syntactic restrictions are imposed (as in the possessive) to compensate for loss of suffixes. Again, this has characterized the development of English generally. We can see this still in operation in standard English possessives. Although *the man's leg* is a perfectly natural way of expressing possession, *the table's leg* seems odd. We are more likely to say *the table leg* or *the leg of the table*. In the first, the word order tells us the relationship between the two nouns; in the second, a prepositional phrase is created to express the relationship. Under what conditions are we more likely to use the suffix as opposed to word order restrictions or prepositional phrases to express possession?

14. How do you think the development of computer spelling checkers will affect our attitudes towards standard spelling?

15. Noah Webster was instrumental in developing an American standard of spelling as distinct from the British standard. What are the British spellings of these words: *anemic, curb, center, labor, unionize*?

16. Alphabets, where each symbol (ideally) represents one phoneme, are only one kind of writing system. Some languages use *syllabaries*, where each symbol represents a whole syllable. For example, a word like *sofa* would be written with two symbols, one that stood for *so* and one that stood for *fa*. Do you think a syllabary would be a good writing system for English?

17. Another possible writing system is a word-writing system, or an ideographic system. In such systems each symbol represents a meaning, not a sound or a combination of sounds. Chinese uses an ideographic system. It is especially useful for Chinese because spoken Chinese is made up of many mutually unintelligible dialects. Speakers of these dialects can, however, share the same writing system because the symbols refer directly to meaning. Written English uses some word-writing symbols: &, $, @, for example. Do you think a word-writing system would be a good system for English generally?

18. Sometimes the spelling of a word can lead to a change in the pronunciation of the word. For example, many people pronounce the *t* in *often* because of its spelling. Can you think of other words in which people have come to pronounce silent letters?

19. Sometimes spelling is deliberately distorted. What is the effect of quoting a person as saying "Who sez she wuz angry?" Does this represent a different pronunciation from "Who says she was angry?" This distortion for effect is called *eye dialect*. Can you think of other examples?

20. One current trend in American English pronunciation is to preserve the pronunciation of a root morpheme when it combines with other morphemes. For example, *compárable* is a more recent pronunciation of *cómparable*. Can you think of other examples of this change in pronunciation?

PRACTICE EXERCISES (Answers on p. 281) _____

1. Divide the following words up into their component morphemes. Tell the meaning or grammatical function of each morpheme: *sportsmanlike, neighborhood, irreversible, anti-abortion, revitalization*

2. Which are the roots and which are the affixes in the above words?

3. Give the phonetic description of each of these consonants: *b, t, g, n, v, s*

4. Which of these sounds are voiced? *p, s, z, g, m, d, k*

5. Which of these sounds are stops? *b, z, k, t, p, f, v*

6. What sound results if you remove vocal cord vibration from: *b, z, g, v, d*?

7. Tell whether each word begins with a stop, a fricative, or an affricate: *cheap, zany, pull, feisty, think, jump, threw, very, bulb, gone*

8. How is the present tense suffix expressed *phonetically* for each of the following verbs: *run, hit, catch, kiss, kill*

9. How is the past tense suffix expressed *phonetically* for each of the following verbs: *talk, pat, rub, toss, raid*

10. How is the possessive suffix expressed *phonetically* for each of the following nouns? *Jane's, Charles's, Mike's, Biff's, Pearl's*

11. What strategy is being used by a child who says *aminal* for *animal*?

12. In which of the following words would a speaker of an *r*-less dialect of English pronounce the *r*: *card, receipt, around, sister, bran?65+*

13. What would be an AAVE version of the following sentences:

    1. Does the boy's father need ten dollars?

    2. She wants to meet us at Mary's aunt's house.

    3. The child's happy because he found five cents.

14. List ten inconsistencies of the English spelling system illustrated by the following words: *write, right, rite, wine, my, sigh, sit, cynic, cane, canny, yeast, see, lien*

ANSWERS
TO PRACTICE EXERCISES

CHAPTER 3

1. *Note:* Inanimate nouns are nonhuman; abstract nouns are inanimate.
   1. *calcium:* common, inanimate, noncount, concrete
      element: common, inanimate, count, concrete
      bones: common, inanimate, count, concrete
   2. *instructor:* common, animate, human, count, concrete
      students: common, animate, human, count, concrete
      exercises: common, inanimate, count, concrete
   3. *worker:* common, animate, human, count, concrete
      factory: common, inanimate, count, concrete
      dissatisfaction: common, noncount, abstract
      colleagues: common, animate, human, count, concrete
   4. *cats:* common, animate, nonhuman, count, concrete
      dogs: common, animate, nonhuman, count, concrete
      friendship: common, noncount, abstract
      love: common, noncount, abstract
      people: common, animate, human, count (plural only), concrete
   5. *animal:* common, animate, nonhuman, count, concrete
      Minnesota: proper, inanimate, count, concrete
      gopher: common, animate, nonhuman, count, concrete

2. 1. *wind*—count; *hole*—count, *tent*—count
   2. *sugar*—noncount, *diet*—count, *health*—noncount
   3. *mother*—count, *wine*—noncount, *dinner*—count
   4. *dinner*—noncount, *meal*—count, *day*—count
   5. *babysitter*—count, *baby*—count, *milk*—noncount, *peas*—count

3. 1. *both my older sisters*
      pre det   head
   2. *half of his geometry class*
      pre det   head
   3. *what an exciting event*
      pre det   head
   4. *the girl's third attempt*   and *the girl's*
      det   post head   det head
   5. *three boats*
      det head

6. *my seven cousins*
   det post head
7. *such a shame*
   pre det head
8. *all the children*
   pre det head

4. 1. *my:* possessive pronoun
   2. *his:* possessive pronoun
   3. *an:* indefinite article
   4. *the girl's:* possessive noun phrase and *the:* definite article
   5. *three:* quantity
   6. *my:* possessive pronoun
   7. *a:* indefinite article
   8. *the:* definite article

5. 1. *the weekend:* subject
      my favorite part of the week: subject complement
      the week: object of a preposition
   2. *Martha's boss:* subject
      Martha's: determiner
      a bonus: direct object
   3. *Bob:* subject
      the doctor: direct object
      the answer to his prayers: object complement
      his prayers: object of a preposition
   4. *a quaint old boat:* subject
      the river: object of a preposition
      Sally's farm: object of a preposition
      Sally's: determiner
   5. *the cashier:* subject
      the receipt: direct object
      the woman: indirect object
      the fur coat: object of a preposition

6. 1. subject*: the child*
      direct object: *a present*
      indirect object: *her mother*
      object of a preposition: *Mother's Day*
   2. subject: *the first day of April*
      subject complement: *my favorite day of the year*
      object of a preposition: *April*
      object of a preposition: *the year*

7. All have indirect objects. 1., 4., and 5. are inverted.

8. 1. *Mr. Allen   taught   the boy   geometry.*
        subject   indirect object   direct object

    2. *Mr. Allen   considered   the boy   an idiot.*
       subject   direct object   object complement

9.   1. *winning a race:* gerundive subject
      2. *working hard:* gerundive object of a preposition
      3. *to hike in the woods:* infinitival direct object
      4. *to live honestly:* infinitival subject
      5. *not trying:* gerundive object of a preposition
      6. *to remain calm:* infinitival subject complement
      7. *all this whining:* gerundive direct object
      8. *learning a new skill:* gerundive direct object
      9. *waiting for the results:* gerundive subject complement
    10. *to arrive tomorrow:* infinitival direct object

10.   [[*the young woman's*] *father*]
        [*the farm* [*at the foot of* [*the hill*]]]

CHAPTER 4

1.   1. *to ask:* infinitive
      2. *looking:* present participle
      3. *considered:* past participle
      4. *be:* infinitive
         expecting: present participle
      5. *having:* present participle
         been: past participle
         informed: past participle
      6. *driving:* present participle
      7. *to be:* infinitive
         like: infinitive
      8. *leave:* infinitive
      9. *sitting:* present participle
         see: infinitive
    10. *have:* infinitive
         been: past participle
         living: present participle

2.   1. gerund
      2. present participle
      3. present participle
      4. present participle
      5. gerund—gerund
      6. present participle
      7. gerund
      8. gerund

    9. present participle

   10. gerund

3.   N = nonstandard, S = standard

    1. N

    2. N

    3. N

    4. S

    5. S

    6. S

    7. N

    8. S

    9. S

   10. N

4.   1. *can:* helping, modal
       leave: main

    2. *are:* main

    3. *is:* helping, auxiliary
       running: main

    4. *has:* main

    5. *have:* helping, auxiliary
       been: helping, auxiliary
       expecting: main

    6. *would:* helping, modal
       wait: main

    7. *has:* helping, auxiliary
       had: main

    8. *has:* helping, auxiliary
       been: helping, auxiliary
       having: main

    9. *must:* helping, modal
       stop: main

   10. *must:* helping, modal
       leave: main

5.   simple present: *Mary goes home.*
     simple past: *Mary went home.*
     simple future: *Mary will go home.*
     present progressive: *Mary is going home.*
     past progressive: *Mary was going home.*
     future progressive: *Mary will be going home.*
     present perfect: *Mary has gone home.*
     past perfect: *Mary had gone home.*
     future perfect: *Mary will have gone home.*
     present perfect progressive: *Mary has been going home.*

past perfect progressive: *Mary had been going home.*
future perfect progressive: *Mary will have been going home.*

6. 1. simple past
2. future perfect progressive
3. simple present
4. present perfect
5. past perfect
6. simple future
7. past progressive
8. present perfect progressive
9. future perfect
10. past perfect progressive

7. 1. *flew:* intransitive
2. *flew:* transitive
3. *seemed:* linking
4. *appeared:* intransitive
5. *felt:* linking
6. *saw:* transitive
7. *was:* linking
8. *sobbed:* intransitive
9. *landed:* intransitive
10. *cheered:* intransitive

8. 3. and 4. have dangling participles.

9. 2., 3., 4., 6., 9., and 10. are violations of subject-verb agreement.

10. Examples:
1. *You may help yourself to a cookie.*
2. *I am tired.*
3. *She felt the tension in the room.*
4. *If I had the time, I could build it myself.*
5. *He must be there by now.*
6. *Will you listen for just a minute?*
7. *They looked furious.*
8. *They looked high and low.*
9. *I tasted the stew.*
10. *The man bent to help the child.*

CHAPTER 5

1. 1. *his*/Max's, *it*/his book
2. *we*/speaker + other(s)

 3. *her*/Alice
 4. *you*/addressee(s), *it*/the exercise
 5. *my*/speaker
 6. *I*/speaker, them/the classes
 7. *she*/Sally, *him*/Joe
 8. *you*/addressee(s), *it*/the missing letter, me/speaker
 9. *he*/my uncle, *my*/speaker, *me*/speaker
 10. *you*/addressee(s), *us*/speaker + other(s)

2. 1. I
 2. them
 3. your
 4. she
 5. him
 6. us
 7. you
 8. its
 9. our
 10. they

3. 1. mine
 2. yours
 3. theirs
 4. his
 5. hers

4. N = nonstandard, S = standard
 1. S
 2. N
 3. N
 4. N
 5. N
 6. S
 7. S
 8. N
 9. N
 10. S

5. 1. ourselves
 2. herself
 3. themselves
 4. yourselves
 5. myself

6.   1. subject and object refer to same entity
     2. contrast
     3. alone
     4. contrast
     5. subject and object refer to same entity

7.   1. who
     2. whom
     3. whom
     4. who
     5. whom
     6. who
     7. who
     8. whom
     9. who
   10. who

8.   1. what
     2. who
     3. which
     4. whom
     5. whose

9.   1. universal
     2. reciprocal
     3. relative
     4. demonstrative
     5. indefinite
     6. reciprocal
     7. indefinite
     8. universal
     9. relative
   10. indefinite

10.   1. *somebody* (indefinite human subject)
      me (personal first-person-singular direct object)
    2. *what* (interrogative nonhuman direct object)
      I (personal first-person-singular subject)
      you (personal second-person direct object)
    3. *whose* (interrogative human possessive subject complement)
    4. *everyone* (universal human subject)
      my (personal possessive first-person-singular determiner)

5. *she* (personal third-person-singular feminine subject)
   that (relative direct object)
   they (personal third-person-plural subject)

CHAPTER 6

1. *spectacular:* more spectacular, most spectacular
   rigid: more rigid, most rigid
   mossy: mossier, mossiest
   verifiable: nongradable
   bland: blander, blandest

2. illiterate, unacceptable, unfortunate, irreversible, immobile, dysfunctional, disconnected, impolite, indistinct

3. 1. *strong:* gradable, attributive
       weaker: nongradable, attributive
    2. *only:* nongradable, attributive
       depressed: gradable, predicate
       favorite: nongradable, attributive
    3. *incredible:* gradable, attributive
       stable: gradable, predicate
    4. *angry:* gradable, attributive
       ruined: nongradable, attributive
    5. *sour:* gradable, predicate
    6. *circular, square, rectangular:* nongradable, predicate
    7. *dear:* gradable, attributive
       little: gradable, attributive
    8. *elderly:* gradable, attributive
       incompetent: gradable, predicate
       wrong: nongradable, predicate
    9. *thoughtful:* gradable, attributive
       happy: gradable, predicate
   10. *incapable (of change):* nongradable, predicate

4. 1. *very nervous:* subject complement
    2. *about to explode:* subject complement
    3. *extremely lazy:* object complement
    4. *pleased to inform you of the results:* subject complement
    5. *big colorful leafy:* attributive
    6. *inclined to tell the truth:* subject complement
    7. *highly impertinent:* object complement
    8. *tiny little:* attributive
       brand new: attributive

        9. *very generous:* attributive
          overly needy: attributive
     10. *old battered:* attributive

5.   1. *quite*/adverb; *responsibly*/verb
      2. *nevertheless*/sentence
      3. *extremely*/adjective
      4. *too*/ adverb; *late*/verb
      5. *yesterday*/verb; *quite*/adverb; *well*/verb
      6. *therefore*/sentence
      7. *so*/adverb; *patiently*/verb
      8. *quite*/adjective
      9. *happily*/verb
    10. *however*/sentence

6.   1. *good*
      2. *slow, slowly*
      3. *bad* (*badly,* if *feel* is an action verb)
      4. *loud, loudly*
      5. *well*
      6. *good, well* (different meanings)
      7. *well* (*good,* if he did good things)
      8. *hard*
      9. *soft*
    10. *well, good* (different meanings)

7.   1. quite impudently
      2. loud enough
      3. very easily
      4. rather tentatively
      5. better than her brother
      6. well for her age
      7. as civilly as I could
      8. surprising well
      9. as quickly as possible
    10. extraordinarily fast

CHAPTER 7

1.   1. the springtime
      2. the seashore
      3. you and me
      4. no obligation

     5. hook, crook

     6. us

     7. my birthday

     8. the right moment

     9. class

   10. working hard

2.   1. with the flea collar: adjectival

      2. in the afternoon: adverbial

      3. in the red helmet: adjectival

      4. on the counter: adjectival

      5. under the tree: adjectival

      6. of worms: adjectival

      7. into the board: adverbial

      8. by tomorrow: adverbial

      9. before the judge: adverbial

    10. at the dry cleaners: adjectival

3.   Meaning 1: *I used a radar detector to find the child*
           The prepositional phrase is adverbial.

      Meaning 2: *I found the child who had the radar detector*
           The prepositional phrase is adjectival.

4.   into a pipe; under the house; around the corner
      one one reading: [into a pipe [under the house [around the corner]]]

5.   1. For whom are you looking?

      2. That is the book about which she was talking.

      3. Is this the pot in which I am supposed to cook it?

      4. I need to call the friends with whom I am going.

      5. For which actor are you standing in?

      6. I fear the world in which we live.

      7. Is this the channel on which the news is broadcast?

      8. On which bench did they sit?

      9. This is the hill down which we rolled as children.

    10. Off which cliff did she jump?

6.   1. She turned the ignition off.

      2. Linda took the trash out.

      3. The teacher brought the books in.

      4. The librarian looked the address up for me.

      5. Lois put her new dress on.

      6. Arsonists burned the building down.

      7. The children turned their parents in to the police.

      8. Please hand your homework in.

    9. They turned our offer down.
  10. Hand your money over!

7.   1. particle
     2. either
     3. preposition
     4. particle
     5. particle
     6. particle
     7. preposition
     8. particle
     9. preposition
   10. preposition

8.   2., 4., 6., 7., 9., 10., because the object of the preposition is a pronoun.

9.   1. adverb
     2. particle
     3. preposition
     4. preposition
     5. adverb
     6. particle
     7. particle
     8. preposition
     9. adverb

CHAPTER 8

1.   1. verb, noun, adjective
     2. adjective, verb, noun
     3. verb, adjective, noun

2.   1. preposition
     2. adverb
     3. particle
     4. adjective
     5. verb
     6. adjective

3.   1. adjective, present participle
     2. gerund
     3. present participle
     4. adjective
     5. gerund
     6. gerund, adjective

7. present participle
8. *buying:* present participle; *skiing:* adjective
9. adjective
10. gerund

4. 1. verb, adjective, verb (past participle)
2. verb (past participle), verb, adjective

5. 1. modification
2. modification
3. grouping
4. grammatical relations
5. modification
6. modification
7. modification
8. grammatical relations
9. modification
10. grouping

6. In the first, *more* can be requesting "more pictures" or it can form the comparative of *beautiful*. In the second, *less* forms the negative comparative of *beautiful*, but it cannot (in formal standard English) refer to the quantity of pictures, because *pictures* is a count noun.

7. 1. proper noun (or noun phrase)
2. preposition
3. adjective
4. adverb
5. noun phrase
6. clause
7. pronoun
8. verb phrase
9. verb
10. prepositional phrase

8. 2. Let's get together either on or about the fifteenth.
4. Please read this both carefully and slowly.
5. I visited both my grandmother and my grandfather.
8. Ed both dropped off the package and left the room. (?)
9. I both cut and dried the flowers.
10. Look either under the stove or next to the refrigerator.

9. CC = coordinating conjunction; SA = sentence adverb
1. CC: The sky is cloudy, but it won't rain.
2. CC: The class was delayed, for no one had a book.
3. SA: My cousins were ill; nevertheless, they visited me.

    4. CC: The roast was ready, so I took it out of the oven.

    5. SA: We had a good time; therefore, we exchanged phone numbers.

    6. SA: Cats are affectionate; moreover, they are loyal.

    7. SA: Our trip to Quebec was tiring; however, we enjoyed it.

    8. CC: The elections were over, and our party had won.

    9. SA: I'm giving you the day off; furthermore, I'm increasing your salary.

    10. CC: Turn yourself in, or I will have to report you.

Alternate punctuation: coordinating conjunctions do not require commas; sentence adverbs may begin a new sentence.

10.
    1. The boy (fished in the pond) and his sister fished in the pond.

    2. Mary baked the potatoes and (Mary) (baked) the squash.

    3. Bill bought a house and his brother (bought) a condo.

    4. The squirrel ran up (the tree) and (the squirrel) (ran) down the tree.

    5. All children need love and (all children) crave attention.

    6. Her teeth are strong and (her teeth) (are) white.

    7. The customer called the company and (the customer) threatened to sue.

    8. All my friends (came to our party) and all her friends came to our party.

    9. Convertibles are nice in summer and hardtops (are nice) in winter.

    10. The man looked in the drawer and the woman (looked) in the cupboard.

11. Repeated prepositions cannot be ellipted.

CHAPTER 9

1.
    1. Our house was sold by the realty company.

    2. Widespread destruction was caused by the storm.

    3. Great sensitivity is required by this job.

    4. An explosion was created by mixing the two chemicals.

    5. A new chancellor is being hired by a consulting firm.

    6. The speech will be delivered tomorrow by our president.

    7. A book was left at the circulation desk by the librarian.

    8. Their cruelty has been witnessed by many people.

    9. A week's worth of groceries was purchased by the family with your gift.

    10. The first ball was thrown by a retired outfielder.

2. 1., 9., and 10. are candidates. Others are possible with some change in meaning. For example, replacing *be* with *get* in 6. suggests that there will be some obstacle in the way of delivering the speech.

3.
    1. *the teller:* subject
       the customer: indirect object
       a roll of bills: direct object

    2. *your proposal:* grammatical subject, logical direct object
       management: grammatical object of a preposition, logical subject

    3. *the stagecoach:* grammatical subject, logical direct object
       the bandits: grammatical object of a preposition, logical subject
    4. *that unhappy child:* subject
       constant attention: direct object
    5. *the customer:* grammatical subject, logical indirect object
       a roll of bills: direct object
       the teller: grammatical object of a preposition, logical subject
    6. *the book:* grammatical subject, logical direct object
       the library: object of a preposition
       the sheepish patron: grammatical object of a preposition, logical subject
    7. *your party:* subject
       the fountain: object of a preposition
    8. *many intelligent people:* subject
       good writing: direct object
    9. *the suspect:* grammatical subject, logical direct object
       his rights: object of a preposition
       the police: grammatical object of a preposition, logical subject
  10. *free cheese:* grammatical subject, logical direct object
       the government: grammatical object of a preposition, logical subject

4.   The nervous patient was given a clean bill of health by the doctor.

5.   1. The passive voice should be avoided.
    2. That old tree was chopped down.
    3. You are expected to dress appropriately.
    4. The bill was passed by a vote of 50–37.
    5. The trash is picked up on Tuesdays.
    6. The news is broadcast at 6:00 every evening.
    7. The Yankees were defeated.
    8. The store will be opened early tomorrow.
    9. A warrant was obtained to search the house. (Or, A warrant to search the house was obtained.)
  10. Laser printers are preferred.

6.   The logical subjects can be derived from the rest of the sentence.

7.   1. Lunch should be eaten by employees in the cafeteria.
    2. The animals were released from their cages by the zookeeper.
    3. Her cries for help were ignored by everyone.
    4. The law was repealed by the legislature.
    5. The bell was rung one last time by the bellringer.
    6. My car will be repaired by the mechanic.
    7. The rules of the competition were rarely understood by the participants.
    8. Our mail is being sorted right now by the postal worker.
    9. Her gardens were planted in May by her gardener.
  10. The students were told by the teacher to begin the exam.

8. 1. the child
    2. the clown
    3. a handful of coins
    4. the child
    5. coins, the clown

9. A handful of coins was given to the child by the clown.

10. 1., 2., 5., 6., and 10. are ambiguous. Each can be analyzed as a linking verb + adjective or as a truncated passive.

CHAPTER 10

1. 1. exclamative (!)
    2. interrogative (?)
    3. imperative (!)
    4. declarative (.)
    5. imperative (.)
    6. exclamative (!)
    7. interrogative (?)
    8. interrogative (?)
    9. declarative (.)
    10. imperative (!)

2. 1. Do we need to rotate the tires?
    2. Is help on the way?
    3. Can you give me an estimate on the costs?
    4. Did he do the assignment before class?
    5. Did the laborers rest after lunch?
    6. Does word processing save us a lot of time?
    7. Must we observe the rules?
    8. Are they aware of the danger?
    9. Did I spell the word wrong?
    10. Won't this dog fetch?

3. 1. Whose book did you borrow?
    2. Who needs a ride?
    3. What can they do for me?
    4. Where can I find good corned beef?
    5. For whom was she asking?
    6. What fell on my head?
    7. Which did she order?
    8. Why did he remain silent?

9. With whom do you wish to speak?

10. Whom did the scandal destroy?

4. 1., 7., 8., 9., and 10. require the insertion of *do*. 2. and 6. question subjects, and the remainder have helping verbs.

5. 1. didn't it?
   2. don't they?
   3. can we?
   4. isn't it?
   5. didn't they?
   6. didn't you?
   7. haven't they?
   8. don't I?
   9. hasn't he?
   10. doesn't she?

6. 1., 2., 5., 6., 8., and 10. require insertion of do, because the statement contains no helping verb.

7. 1. *Barely* is only partially negative so it is unclear how to reverse it.
   2. *No one* has no matching gender-neutral singular pronoun.
   3. *Someone:* same as for *no one.*
   4. The gender of the baby may be unknown to the observer and *it* is only for nonhumans.
   5. *Hardly* is only partially negative (same as 1.)

8. 1. yes-no echo
   2. embedded
   3. wh-
   4. tag
   5. wh-
   6. wh- echo
   7. yes-no
   8. tag
   9. declarative with the force of a question
   10. wh-

9. 1. What a wonderful parent Al is!
   2. How ugly that building is!
   3. How my heart aches for you!
   4. What a cute baby they have!
   5. How bitterly she complained!
   6. How awful this is for you!
   7. How he boasts about his children!

8. How well Senta performed in the recital!
9. How proud of her Kara and Jan were!
10. How tiresome this discussion is!

10. Examples:

1. The dog needs to be walked; Can you take the dog for a walk?
2. I need you to pick up some milk on your way home; Could you pick up some milk on your way home?
3. I want you to talk to me about your concerns; Won't you talk to me about your concerns?
4. I need help with these groceries; Could you help me with these groceries?
5. I suggest that you clean up this mess; I wonder if you might clean up this mess.
6. You must look for a job; Will you look for a job?
7. You ought to rethink your demands; Might you rethink your demands?
8. I'd like you to host a reception for them; Would you host a reception for them?
9. You might consider your alternatives; Won't you consider your alternatives?
10. It would be wise to get there on time; Can you get there on time?

CHAPTER 11

1.
1. You shouldn't speak ill of the dead.
2. Karel can't cope with the situation.
3. _____
4. Sal isn't listening to you.
5. We weren't expecting you.
6. I'm not surprised.
7. We shan't be daunted.
8. They mustn't think we're ungrateful.
9. He won't be allowed to perform.
10. That wouldn't help me.

Exceptions: *Mayn't* isn't used. There is no form for *am not* (*I'm* is a contraction of *I* and *am*). *Won't* is an irregular form for *will not*. *Shan't* is rarely, if ever, used in American English.

2. 2., 5., 6., 7., and 10. require *do* because they contain no helping verb.

3.
1. The printer has no ink; The printer does not have any ink.
2. No reasonable person would tolerate this behavior; A reasonable person would not tolerate this behavior.
3. I have no idea; I do not have any idea.

4. She received no compensation for the job; She didn't receive any compensation for the job.

5. He got no love from his grandparents; He didn't get any love from his grandparents.

4.   N = nonstandard, S = standard
   1. N
   2. N
   3. S
   4. S
   5. N
   6. S
   7. N
   8. N
   9. S
   10. S

5.   1. I see no people.
   2. Does no proposal suit you?
   3. She will accept no cash.
   4. There is no reason to stay.
   5. Why were there no police at the event?

6.   *unhappily, indecent, disproportionately, inadvisable, impolitely, irrelevant, illiterate, unsavory, nontoxic, unremarkably*

7.   1. We are neither proud nor arrogant.
   2. I neither cleaned the garage nor swept the driveway.
   3. She spoke neither clearly nor accurately.
   4. They will tolerate neither your laziness nor your impudence.
   5. Neither smoking nor drinking is good for you.

8.   Examples:
   1. I seldom go to the movies during the week.
      Seldom do I go to the movies during the week.
   2. Marie rarely studies with Ken.
      Rarely does Marie study with Ken.
   3. We barely arrived on time.
      Barely did we arrive on time.
   4. The farmers were hardly able to support their families.
      Hardly were the farmers able to support their families.(??)
   5. He barely remembers their meeting.
      Barely does he remember their meeting.
   6. The patient could scarcely speak.
      Scarcely could the patient speak (??)

7. There is hardly enough time for this exam.
   Hardly is there enough time for this exam (??)
8. The tenants rarely complained to the landlord.
   Rarely did the tenants complain to the landlord.
9. The beams barely held up the roof.
   Barely did the beams hold up the roof.
10. There is seldom any food in the refrigerator.
   Seldom is there any food in the refrigerator.

9.
1. The food isn't edible; The food is inedible.
2. We didn't go anywhere last night; We went nowhere last night.
3. Our approach isn't confrontational; Our approach is nonconfrontational.
4. There is no reason to worry; There isn't any reason to worry.
5. These reviews aren't spectacular; These reviews are unspectacular.
6. She didn't buy any bonds; She bought no bonds.
7. I'll never read it; I won't read it anytime.
8. There is no paint in the garage; There isn't any paint in the garage.
9. No car can travel this road; A car can't travel this road.
10. That star has no fan clubs; That star doesn't have any fan clubs.

10.
1. active, affirmative, interrogative
2. passive, affirmative, declarative
3. passive, negative, declarative
4. passive, negative, imperative
5. active, affirmative, interrogative
6. active, negative, declarative
7. active, affirmative, imperative
8. passive, negative, interrogative
9. active, negative, declarative
10. active, negative, imperative

CHAPTER 12

1. C = coordination, S = subordination
1. C
2. C
3. S
4. C
5. S
6. C
7. S
8. C

      9. C
     10. S

2.   1. subordinating conjunction
     2. coordinating conjunction
     3. coordinating conjunction
     4. sentence adverb
     5. preposition
     6. subordinating conjunction
     7. sentence adverb
     8. coordinating conjunction
     9. subordinating conjunction
    10. preposition

3.   1. *that you like doing housework:* extraposed subject
     2. *For Karen to quit now:* subject
     3. *That she loves her children:* subject
     4. *that I was busy:* object
     5. *for your associate to redo your work:* extraposed subject
     6. *his friends to entertain him:* object
     7. *that the plan is unworkable:* extraposed subject
     8. *that we can't see you more often:* object
     9. *the animals to sleep a lot:* object
    10. *that she receive the award:* extraposed subject

4.   N = noun clause, A = adverbial clause
     1. A
     2. N
     3. N
     4. A
     5. N
     6. A
     7. N
     8. N
     9. A
    10. A

5.   O = object noun clause, C = complement noun clause
     1. O
     2. C
     3. C
     4. O
     5. C
     6. O

> 7. C
> 8. C
> 9. O
> 10. O

6. 1. that is located in my neighborhood; the branch library
   2. to whom you were speaking; the person
   3. whose patients complained; the doctor
   4. that I mentioned; the painting
   5. who understands my problem; someone
   6. that is crying; the baby
   7. that I gave you; one example
   8. whom she had accused; the man
   9. which houses the pandas from China; the zoo
   10. who adopted the twins; the couple

7. 1. *since you're already here:* adverbial clause
   2. *Bill to play soccer:* object noun clause
   3. *that you refrain from smoking:* extraposed subject noun clause
   4. *that I can't go home until spring:* subject complement noun clause
   5. *that she admires you:* subject noun clause
   6. *that she would support our decision:* object noun clause
   7. *if you have the time:* adverbial
   8. *who accosted you:* relative clause
   9. *that my grandparents told me:* relative clause
   10. *for Chris to lie about her age:* subject noun clause

8. 1., 3., 4., and 5. could be nonrestrictive if the head were fully identified without the clause. In 1., for example, the relative clause could be made nonrestrictive if *my cousin* had already been mentioned in the conversation.

9. 1. whiz-deletion
   2. omit the object relative pronoun
   3. no reduction because it's nonrestrictive
   4. whiz-deletion
   5. whiz-deletion
   6. no reduction because the relative pronoun is a subject
   7. whiz-deletion
   8. whiz-deletion
   9. omit the object relative pronoun
   10. no reduction because it's nonrestrictive

10. 1. *standing on the corner:* nonfinite verb phrase with a present participle
   2. *insulted by the remark:* nonfinite verb phrase with a past participle
   3. *in the park:* prepositional phrase

4. *expecting a call:* nonfinite verb phrase with a present participle
5. *made of straw:* nonfinite verb phrase with a past participle

Note: Nonfinite verb phrases with participles are sometimes called *participial phrases.*

CHAPTER 13

1. 1.

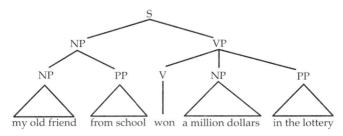

2.

the curve   in the road   caused   an accident   on that hill

3.

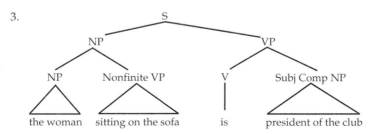

2. These are all compound sentences.
   Examples:
   (a) The weather was bad but we enjoyed ourselves anyway.
   (b) Either the plan will fail or it will revolutionize the industry.
   (c) Lynn loved her work and she needed the money, yet she couldn't stay in Ohio.

3. These are all complex sentences:
   1. *that you never smile:* noun
      that you are sad: noun
   2. *that the clerk showed me:* relative

3. *who are rich:* relative
   who are poor: relative
4. *whose mother died:* relative
   that he wanted to live with his uncle: noun
5. *before they arrived:* adverbial
6. *whom you recommended:* relative
   because I knew [that she would give me sound advice]: adverbial
   that she would give me sound advice: noun
7. *when you get to town:* adverbial
   that you visit the library [that was just built]: noun
   that was just built: relative
8. *that laughed [when you entered the room]:* relative
   when you entered the room: adverbial
   that you were upset: noun
9. *wherever they can find the best bargains:* adverbial
10. *for Bob to understand [that we meant no harm]:* noun
    that we meant no harm: noun

4.

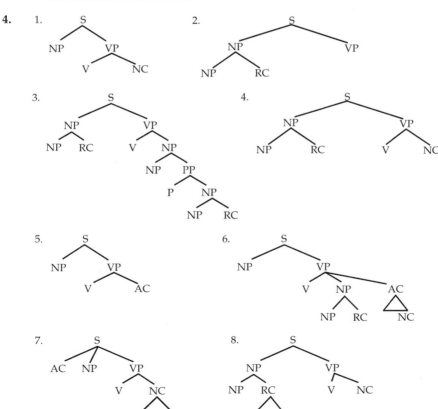

9.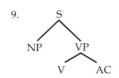

10.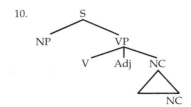

5. Examples:
   (a) That she is angry surprises me.
   (b) The man who left the room said that he would return.
   (c) My sister found the purse that she had lost when she got home.
   (d) When they arrived, the guard who stood at the door said that they couldn't enter.
   (e) That she came to my party suggests that she likes me.

6. (a) The children thought that there would be treats at the party and that they would have fun.
   (b) Jennifer knew that there would be trouble when she got there.
   (c) Tim identified the man who said that the world would end tomorrow.

7. CD = compound, CX = complex
   1. CX
   2. CX
   3. CD
   4. CD
   5. CX
   6. CX
   7. CX
   8. CD
   9. CX
   10. CD

8. These are all compound sentences.
   Examples:
   (a) The woman whom you admire is honest but she is tactless.
   (b) Rick knew that it would rain, yet he suggested that we hold the picnic.
   (c) Everyone came, and the people who arranged the meeting smiled when they saw the crowd.
   (d) The villagers cried when they heard the news, for the peace which they had wished for had finally arrived, yet the negotiators knew that the peace would not last and the fighting would resume.

9.

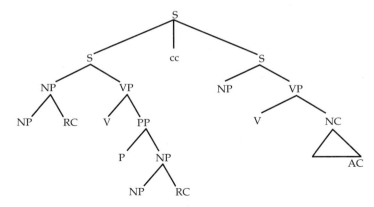

10. S = simple; CD = compound, CX = complex, CC = compound-complex
1. S
2. CX
3. CX
4. CX
5. CC
6. S
7. CD
8. CX
9. CD
10. CX

CHAPTER 14

1. (obvious meanings omitted)

sport	*neighbor*	*ir:* not	*anti:* against	*re:* again
s: possessive	*hood:* domain	*re:* back	*abort*	*vit:* life
man		*verse:* turn	*ion:* noun	*al:* adjective
like		*ible:* adjective		*ize:* verb
				ation: noun

2. roots: *sportsman* (compound)
  neighbor
  verse
  abort
  vit

prefixes: *ir, anti, re*
suffixes: *like, hood, ible, ion, ize, ation*

3. *irregular, irrelevant, irresistible*
   illegal, illegitimate, illegible

4. *b:* oral, voiced, stop, two lips together
   t: oral, voiceless, stop, tongue on alveolar ridge
   g: oral, voiced, stop, back of tongue on velum (soft palate)
   n: nasal, voiced, tongue on alveolar ridge
   v: oral, voiced, fricative, upper teeth on lower lip
   s: oral, voiceless, fricative, tongue on alveolar ridge

5. *z, g, m, d*

6. *b, k, t, p*

7. *b- > p; z- > s; g- > k; v- > f; d- > t*

8. | | |
   |---|---|
   | *cheap:* affricate | *jump:* affricate |
   | *zany:* fricative | *threw:* fricative |
   | *pull:* stop | *very:* fricative |
   | *feisty:* fricative | *bulb:* stop |
   | *think:* fricative | *gone:* stop |

9. *run: z*
   hit: s
   catch: vowel + *z*
   kiss: vowel + *z*
   kill: z

10. *talk: t*
    pat: vowel + *d*
    rub: d
    toss: t
    raid: vowel + *z*

11. *Jane's: z*
    Charles's: vowel + *z*
    Mike's: s
    Biff's: s
    Pearl's: z

12. metathesis

13. receipt, around, bran

14. 1. Do the boy father need ten dollar?
    2. She want to meet us at Mary aunt house.
    3. The child happy because he found five cent.

15. Examples:
    1. The *w* in *write* is silent but it is pronounced in *wine*.
    2. *write, right,* and *rite* are pronounced the same but spelled differently.

3. The *i* sound is represented by the letter *i* in *right* but *y* in *my*.
4. The *s* sound is *s* in *sit* but *c* in *cynic*.
5. The vowel of *sit* is *i*, but the same sound in *cynic* is represented by *y*.
6. The *gh* in *right* and *sigh* is silent.
7. The *a*'s in *cane* and *canny* represent two different sounds.
8. The *ea* of *yeast*, the *ee* of *see*, and the *ie* of *lien* represent the same sound.
9. The *y* is a vowel in *canny* but a consonant in *yeast*.
10. The letter *c* has an *s* sound in *cynic* but a *k* sound in *cane* and *canny*.

GLOSSARY

Note: Numbers in parentheses indicate the chapters in which the term is either introduced or discussed.

abstract noun: a subcategory of noun that refers to ideas or concepts (**3**)

active voice: sentence structure that follows the order *doer + action + receiver* (**9**)

adjective: the lexical category that typically modifies nouns (**6**)

adjective complement: an element following an adjective in an adjective phrase that completes the meaning of the adjective (**6**)

adjective negation: expressing negation by attaching a negative prefix to an adjective (**11**)

adjective phrase: a phrase with an adjective as its head (**6**)

adverb: the lexical category that typically modifies verbs, adjectives, or other adverbs (**6**)

adverb complement: an element following an adverb in an adverb phrase that completes the meaning of the adverb (**6**)

adverb negation: expressing negation by attaching a negative prefix to an adverb (**11**)

adverb phrase: a phrase with an adverb as its head (**6**)

adverbial clause: a subordinate clause that plays the role of an adverb (**12**)

affirmative: describes a positive clause, one which contains no markers for negation (**11**)

affix: a bound morpheme attached to a root (**2, 14**)

affricate: a speech sound that begins as a stop but is released as a fricative (**14**)

African American Vernacular English: a dialect of American English (**14**)

allomorphs: variations of the same morpheme (**14**)

allophones: variations of the same phoneme (**14**)

alphabet: a writing system in which each symbol stands for a phoneme (**14**)

ambiguity: more than one interpretation of a grammatical structure (**8**)

animate noun: a subcategory of noun that refers to animals and humans (**3**)

antecedent: the noun phrase that a pronoun stands for (**5**)

approximant: phonetically, a class of sounds between consonants and vowels in terms of degree of air obstruction (**14**)

article: a type of determiner; see **definite article** and **indefinite article** (**3**)

aspect: a feature added to the time of an action to show its relationship to another time (**4**)

assimilation: the process whereby two sounds are made more alike (**14**)

attributive adjective: an adjective within the same noun phrase as the noun it modifies (**6**)

auxiliary verb: a type of helping verb that carries no meaning of its own (**4**)

bare infinitive: the infinitive without *to*; also called the **base form** of the verb (**4**)

base form: see **bare infinitive**

bound morpheme: a morpheme that must be attached to another morpheme; also called an **affix** (**14**)

case: the grammatical function of pronouns: subject, object, or possessive (**5**)

category-shift: moving an element from one lexical category to another; also called **conversion** (**4, 8**)

clause: a combination of a noun phrase (subject) and a finite verb phrase (predicate) (**2, 12**)

collective noun: a noun that is grammatically singular but has plural meaning (**4**)

common noun: a subcategory of noun that refers to general categories (**3**)

comparative: the form of an adjective used to compare two entities (**6**)

complement: a noun phrase or adjective that describes or renames another noun phrase in the same sentence (**3, 6**)

complement noun clause: a noun clause that plays the role of a complement in a sentence (**12**)

complementizer: a term sometimes used to refer to a word that introduces a noun clause (**12**)

complex sentence: a sentence containing at least one subordinate clause (**13**)

complex tenses: time combined with the perfect or the progressive aspect (**4**)

compound: the constituent that results when two or more like constituents are conjoined (**8**); also used to refer to words that have more than one root (**14**)

compound negation: the negation of a compound using *neither . . . nor* (**11**)

compound sentence: a sentence consisting of two or more conjoined clauses (**13**)

compound-complex sentence: a compound sentence that contains at least one subordinate clause (**13**)

compounding: see **conjoining**

concrete noun: a subcategory of noun that refers to tangible or visible entities (**3**)

conjoining: creating a new constituent by joining together two constituents of the same type; also called **coordination** and **compounding** (**8, 12**)

consonant: a speech sound produced with significant obstruction of air (**14**)

constituent: words in a sequence that group together and function as a grammatical unit (**2**)

conversion: see **category-shift**

coordinating conjunction: a lexical category used to link together like constituents (**8**)

coordination: see **conjoining**

correlative: a two-part coordinating conjunction (**8**)

count noun: a subcategory of noun that can be counted directly (**3**)

cross-referencing rule: a grammatical rule that marks a relationship between two constituents (**2**)

dangling modifier: a nonfinite predicate that cannot be appropriately linked to a noun phrase in the rest of the sentence (**4**)

dangling participle: a dangling modifier that contains a past or a present participle (**4**)

declarative: the sentence type typically used to give information (**10**)

declarative question: see **echo question**

deferred preposition: a preposition that has been left behind when its object has been moved elsewhere in the sentence (**7**)

definite article: a determiner used to modify an already identified entity (**3**)

deletion: the process whereby one or more sounds are suppressed in pronunciation (**14**)

demonstrative: a determiner or pronoun that signals location of an entity with respect to the speaker (**3, 5**)

dependent clause: see **subordinate clause**

derivational affix: an affix that changes the lexical category or subcategory of a root (**2, 14**)

determiner: one kind of modifier in a noun phrase (**3**)

dialect: set of characteristics within a language associated with a community of speakers (**14**)

direct object: typically the noun phrase that follows the verb and receives the action (**3**)

discourse function: the communicative purpose of an utterance (**10**)

disjunction: a coordinating conjunction that signals a choice beween (or among) the conjoined elements; also used to describe the conjoined structure itself (**4, 8**)

dissimilation: the process whereby two sounds are made less alike (**14**)

echo question: an interrogative that retains the structure of a preceding declarative statement ; also called **declarative question** (**10**)

ellipsis: omitting repeated material, used in this book specifically to refer to omission in conjoined clauses (**8**)

embedded question: a question inserted within a main clause (**10**)

embedded sentence: see **subordinate clause**

epenthesis: adding a sound to create an acceptable sequence of phonemes (**14**)

exclamative: the sentence type used to express a feeling with added emphasis (**10**)

extraposed subject noun clause: a noun clause that has been moved to the end of its sentence (**12**)

finite verb form: a form of the verb that is marked for tense (**4**)

finite verb phrase: a verb phrase with a finite verb form as head (**4**)

flat adverb: an adverb that has the same form as its corresponding adjective (**6**)

free morpheme: a morpheme that can stand alone as a word (**14**)

fricative: a consonant sound made with partial obstruction of air (**14**)

full passive: a sentence in the passive voice that includes the doer of the action (**9**)

gender: grammatical marking on a pronoun to match the sex of its antecedent: masculine, feminine, or neuter (**5**)

gerund: a verbal noun formed by adding *-ing* to a verb root (**3**)

gerundive phrase: a noun phrase with a gerund as head (**3**)

gradable adjective: an adjective capable of expressing degrees and comparison (**6**)

grammatical relations ambiguity: ambiguity that results when a noun phrase can be interpreted as playing more than one grammatical role (**8**)

grammatical subject: the noun phrase that the verb agrees with (**9**)

grouping ambiguity: ambiguity that results when adjacent words can be grouped in different ways to signal different meanings (**8**)

head: the main or core word in a phrase; in a phrase containing a relative clause, it is the noun phrase which the clause describes (**2, 5**)

helping verb: a verb that supports a main verb, typically by carrying tense (**4**)

human noun: a subcategory of nouns; refers to human beings (**3**)

hypercorrection: nonstandard language that results when people misapply a grammatical rule in order to avoid a grammatical error (**1**)

imperative: the sentence type typically used to give orders (**10**)

inanimate noun: a subcategory of noun that refers to entities that are neither human nor animal (**3**)

indefinite article: the determiner used to introduce an entity into a discussion (**3**)

indefinite negation: negation by use of a negative indefinite word (**11**)

indefinite pronoun: a pronoun that refers to unspecified entities or quantities (**5**)

independent clause: see **main clause**

indirect object: typically the animate noun phrase that is the beneficiary of an action (**3**)

indirect object inversion: placing the indirect object noun phrase before the direct object (**3**)

infinitival phrase: a noun phrase with an infinitive as head (**3**)

infinitive: the base form of the verb, or the base form preceded by *to* (**3**)

inflectional affix: an affix that conveys grammatical information and does not change the lexical category of the root (**2, 14**)

intensifier: a type of adverb used to modify an adjective or another adverb (**6**)

interrogative: the sentence type typically used to request information (**10**)

interrogative pronoun: a pronoun used to elicit the identity of an unknown noun phrase (**5**)

intransitive verb: a verb that can stand alone in its verb phrase (**4**)

lexical ambiguity: multiple meanings that result from individual words having more than one meaning (**8**)

lexical category: class of words that have similar grammatical functions and forms, also known as **word-class** and **part of speech** (**2**)

linguistic ambiguity: see **ambiguity**

linking verb: a verb that connects a subject to a description of the subject (**4**)

logical direct object: the receiver of the action (**9**)

logical subject: the doer of the action (**9**)

main clause: the clause that expresses the main idea of the sentence and may contain clauses within it; also called **matrix clause (12)**, **independent clause** (**2**), and **superordinate clause (12)**

main verb: the verb of a clause that expresses an action or a mental state (**4**)

mass noun: see **noncount noun**

matrix clause: see **main clause**

metathesis: the process whereby sounds reverse positions (**14**)

modal auxiliary: see **modal verb**

modal verb: a type of helping verb that adds additional meaning to the main verb; also called a **modal auxiliary** (**4**)

modification ambiguity: ambiguity that results when a modifier can be interpreted as describing more than one constituent (**8**)

modifier: an element in a phrase that describes or limits the head (**2**)

morpheme: an individual piece of meaning or grammatical function in a word (**14**)

morphology: the study of how words are constructed from morphemes (**14**)

nasal: a speech sound made with air exiting through the nose (**14**)

negative: describes a clause that contains a marker for negation (**11**)

noncount noun: a subcategory of noun that cannot be counted directly, also called **mass noun**. (**3**)

nonfinite verb form: a form of the verb that carries no tense of its own (**4**)

nonfinite verb phrase: a verb phrase with a nonfinite verb form as head (**4**)

nongradable adjective: an adjective with absolute meaning; has no degrees (**6**)

nonhuman noun: subcategory of nouns; refers to entities that are not human (**3**)

nonrestrictive relative clause: a relative clause that gives added information about an already identified head (**12**)

noun: the lexical category that names entities (**3**)

noun clause: a subordinate clause that plays the role of a noun phrase, as subject, object, or complement (**12**)

noun negation: expressing negation by associating a negative word with a noun (**11**)

noun phrase: a phrase with a noun as its head (**3**)

number: the grammatical marking of quantity; in English the two numbers are singular (one) and plural (more than one) (**3, 4**)

object complement: a noun phrase or predicate adjective that describes the direct object of a sentence (**3, 6**)

object noun clause: a noun clause that plays the role of direct object in a sentence (**12**)

object of a preposition: the noun phrase that follows a preposition in a prepositional phrase (**3**)

objective case: the case of pronouns used as objects (**5**)

oral: a speech sound made with no air exiting through the nose (**14**)

paraphrase: another way of saying a grammatical structure to reveal its ambiguity (**8**)

part of speech: see **lexical category**

partial negation: effect of using one of the set of partially negative adverbs (**11**)

particle: the second element of a two-part transitive verb that can occur either before or after the direct object (**7**)

particle-movement: the grammatical rule that moves a particle behind the direct object (**7**)

passive test: rearranging a sentence from active to passive voice to determine if a noun phrase is a direct object (**3**)

passive voice: sentence structure that follows the order *receiver + action + doer* (**9**)

past participle: a nonfinite verb form that typically consists of *-ed* or *-en* added to the base form; used in the perfect tenses (**4**)

perfect aspect: a feature added to time to show that the action relates to a later time (**4**)

perfect tense: a verb tense that includes the perfect aspect (**4**)

person: the grammatical feature that refers to the role of the participant in a conversation: speaker (first person), listener (second person), or entity spoken about (third person) (**4**)

personal pronoun: a pronoun that plays the role of subject, object, or possessive (**5**)

phoneme: the abstract idea of a particular speech sound in a language (**14**)

phonetics: the study of the production and perception of speech sounds (**14**)

phonology: the study of how individual languages organize and use speech sounds (**14**)

phrase: a constituent consisting of a single word (the "head" of the phrase) and all of its modifiers (**2**)

place of articulation: the point in the mouth where two organs form an obstruction to produce a speech sound (**14**)

possession: a grammatical marking on nouns and pronouns to indicate ownership and a variety of related concepts (**3**)

possessive case: the case of nouns or pronouns used as possessives (**5**)

possessive noun phrase: a noun phrase in which the head noun is in the possessive case (**3**)

possessive pronoun: a pronoun that expresses ownership; has both a short form and a long form (**5**)

postdeterminer: one kind of modifier in a noun phrase (**3**)

predeterminer: one kind of modifier in a noun phrase (**3**)

predicate: the finite verb phrase of a clause (**2**)

predicate adjective: an adjective that is outside the noun phrase of the noun it modifies (**6**)

predicate nominal: see **predicate nominative**

predicate nominative: a noun phrase that plays the grammatical rule of subject complement (**3**)

prefix: an affix that attaches to the beginning of a root (**2**)

preposition: a word that indicates the relationship of a noun phrase to other noun phrases in the same sentence (**7**)

prepositional phrase: a constituent that consists of a preposition followed by a noun phrase (**3, 7**)

present participle: a nonfinite verb form consisting of the base form + *ing;* used in the progressive tenses (**4**)

progressive aspect: a feature added to time to indicate ongoing or background activity (**4**)

progressive tense: a verb that includes the progressive aspect (**4**)

pronoun: a word that typically stands for or takes the place of a noun phrase (**5**)

proper noun: a subcategory of noun that refers to a specific, named entity (**3**)

quantity: one kind of noun modifier in a noun phrase (**3**)

reciprocal pronoun: like a reflexive pronoun in purpose, but used with plural subjects to express mutual activity (**5**)

reduced relative clause: a relative clause that has lost either its object relative pronoun or a relative pronoun and a form of the verb *be* (**12**)

reflexive pronoun: a pronoun that typically is used to avoid repetition of the same noun phrase within a clause (**5**)

relative clause: a subordinate clause that modifies a noun phrase (**5, 12**)

relative pronoun: a pronoun used within a relative clause to replace the second occurrence of the head noun phrase (**5**)

restrictive relative clause: a relative clause that narrows down and identifies the head noun phrase (**12**)

root: that part of a word that carries the core meaning (**2, 14**)

sentence: a grammatical constituent consisting of one or more clauses (**2, 12**)

sentence adverb: an adverb that modifies an entire sentence, injecting commentary from the speaker (**6**)

simple sentence: a sentence consisting of only one clause (**13**)

simple tense: time with no aspect: present, past, and future (**4**)

split infinitive: an infinitive with something in between *to* and the verb (**3**)

standard American English: that form of English expected in public discourse in the United States: in newspapers and magazines, radio and television news broadcasts, textbooks, and public lectures (**1**)

stop: a consonant sound made with complete obstruction of air (**14**)

structural ambiguity: multiple meanings that result from different possible analyses of grammatical structure (**8**)

subject: typically, the noun phrase of a clause that appears before the verb and performs the action (**2, 3**)

subject complement: a noun phrase or adjective that describes the subject of a sentence (**3, 6**)

subject noun clause: a noun clause that plays the role of subject in a sentence (**12**)

subject-verb agreement: the requirement that the subject and the verb of a clause must match in person and number (**4**)

subjective case: the case of pronouns used as subjects or subject complements (**5**)

subordinate clause: a clause that performs a grammatical function within another cause; also called **embedded sentence** and **dependent clause** (**2, 12**)

subordinating conjunction: lexical category that introduces certain subordinate clauses (**12**)

subordination: the process whereby one clause becomes a grammatical part of another (**12**)

suffix: an affix that attaches to the end of a root (**2**)

superlative: the form of an adjective used to compare more than two entities (**6**)

superordinate clause: see **main clause**

syntax: the organization of words, phrases and clauses into sentences; also the study of that organization (**13**)

tag question: an interrogative attached to the end of a declarative statement (**10**)

tense: the time of the action of the verb, or a combination of time and aspect (**4**)

time: the indicator of when the action occurred relative to the time of the utterance: past, present, or future (**4**)

transitive verb: a verb that must be followed by a direct object (**4**)

truncated passive: a sentence in the passive voice that omits the doer of the action (**9**)

universal pronoun: a pronoun that represents an all-inclusive noun phrase or entity (**5**)

verb: the lexical category that typically refers to actions or mental states (**4**)

verb negation: expressing negation by associating a negative word with a verb (**11**)

verbal noun: a noun that is formed from a verb (**3**)

verb phrase: a phrase with a verb as its head and all modifiers of the verb (**4**)

voiced: a speech sound made with vocal cord vibration (**14**)

voiceless: a speech sound made without vocal cord vibration (**14**)

vowel: a speech sound produced with little or no obstruction of air (**14**)

wh- question: a question-type that seeks a specific piece of information (**10**)

whiz-deletion: reduction of a relative clause by omitting a relative pronoun and a form of the verb *be* (**12**)

word: a combination of a root and all its affixes (**2**)

word class: see **lexical category**

yes-no question: a question-type that can be answered "yes" or "no" (**10**)

INDEX

A

Absolutes, 70
Abstract nouns, 24, 284
Academies, language, 5–6, 9, 10
Active voice, 145–46, 284
Adjective complements, 104–5, 284
Adjective negation, 187–88, 284
Adjective phrases, 104–5, 106, 284
Adjectives, 17, 100–104, 284
   attributive, 102–4
   comparative, 100–102, 108
   gradable vs. nongradable, 17, 102, 130,
     192
   modification of, 102
   negation of, 187–88
   predicate, 103–4
   superlative, 100–102, 108
   usage problems, 109–10
Adverb complements, 111, 284
Adverb negation, 187–88, 284
Adverb phrases, 111, 284
Adverbial clauses, 201–2, 203, 284
Adverbs, 17, 105–10, 284
   common properties, 107
   compared to prepositions and particles,
     123
   comparison of, 107–8, 112
   flat, 107–8
   as intensifiers, 102, 106, 284
   as modifiers of sentences, 106, 135–36
   as modifiers of verbs, 105–6
   negation of, 187–88
   partially negtive, 189–90
   placement , 107
   usage problems, 109–10
Affirmative clauses (*see* Clauses)
Affixes, 16, 235–37, 284 (*see also* Bound
   morphemes)
   derivational, 16–17, 235–36
   inflectional, 16, 235–36
   prefixes, 16, 187–88, 237
   suffixes, 16
Affricates, 242, 284

African American Vernacular English,
   247–50, 284
Age of Reason, 5
Ain't, 183, 191
Allomorphs, 237–38, 244–47, 284
Allophones, 239–40, 251, 284
Alphabets, 251–53, 284
   phonemic, 251, 254
   phonetic, 253
Ambiguity, 130–33, 140, 214, 284
   of grammatical relations, 131
   of grouping, 131
   lexical, 132–33
   of modification, 131
   structural, 130–33
The American Heritage Dictionary of the Eng-
   lish Language, 11–12
Animate nouns, 24–25, 285
Antecedents, 77–78, 285
   unspecified gender of, 81, 85
Approximants, 243, 285
Articles, 17, 27–28, 285 (*see also* Determiners)
   definite, 27–28
   indefinite, 27–28
   use of, 28
Aspect, 56–57, 285 (*see also* Progressive
   tenses)
   perfect, 57–58
   progressive, 56
Assimilation, 244, 253, 285
Attributive adjectives, 102–4, 285
Auxiliary verbs, 51–52, 285
   in complex tenses, 56–60
   in interrogatives, 162–63, 165–66, 168–69,
    176
   in negation, 183–84
   in passive clauses, 147–48

B

Bare infinitive, 4, 285
Base form, 285, (*see* Bare infinitive)

Blends, 253
Bound morphemes, 236, 285 (*see also* Morphemes)

C

Cardinal numbers, 30 (*see also* Postdeterminers)
Case, 79, 285
  of interrogative pronouns, 91–92
  objective, 79
  possessive, 79
  of relative pronouns, 90
  subjective, 79
  usage problems, 83–85
Category shift, 128–30, 285 (*see also* Conversion and Crossover functions)
Caxton, William, 4, 9
Chaucer, Geoffrey, 13, 191
Clause coordination, 137–38 (*see also* Conjoining)
Clause ellipsis (*see* Ellipsis)
Clauses, 16, 18, 64, 285
  active vs. passive, 145–55
  affirmative vs. negative, 182–93
  as components of sentences, 191, 193, 220–30
  coordination of, 137–38
  crossover functions of, 174–76
  discourse functions of, 159–77
  ellipsis in, 137–38
  main, 199–215
  subordinate, 199–215
Collective nouns, 68, 285
Commands (*see also* Imperative clauses)
  softened, 53–54, 174–76
Common nouns, 17, 24, 285
Comparatives, 285
  of adjectives, 100–102
  of adverbs, 107, 109
  double, 111
Comparison:
  of adjectives, 100–102, 111–12
  of adverbs, 107, 112
  and pronoun usage, 5, 84, 95
Complement noun clauses, 206–7, 285
Complements, 105, 285 (*see also* Adjectives, predicate and Noun phrases, functions)
  of adjectives, 104–5, 284
  of adverbs, 111, 284
  identifying, 35
  object, 35
  subject, 35
Complementizer, 203, 285
Complex sentences, 224–28, 229, 285

Complex tenses, 56–60, 285
Compound negation, 188–89, 286 (*see also* Negation)
Compound noun phrases, 66, 83
Compound sentences, 222–24, 229, 286 (*see also* Conjoining)
Compound-complex sentences, 228–30, 286
Compounds, 66, 133–37, 285 (*see also* Conjoining)
  negation of, 188–89
Concrete nouns, 24, 286
Conditionals, 53, 54
Conjoining (*see also* Compounds and Sentences, compound)
  of clauses, 137–38, 197–99, 222–24
  of like constituents, 133–36
  and negation, 188–89
  of subordinate clauses, 227, 228–30
Consonant clusters, 239–40, 245–49
Consonants, 240–43, 286 (*see also* Pronunciation)
Constituent structure, 15–19 (*see also* Constituents)
  described with tree diagrams, 221–22
Constituents, 15–19, 286
  compound, 133–37
  creation of, 128
  hierarchical arrangement, 16, 18, 29, 226–27
  identifying, 15–16
  levels of, 16–17
Contractions, 183–84
Contrary-to-fact statements (*see* Hypothetical *if-then* statements)
Conversion, 47, 128, 286 (*see also* Category shift)
Coordinating conjunctions, 286 (*see also* Conjoining)
  correlatives, 134–36, 222
  meanings, 197–99
  punctuation, 140
  simple, 134–36, 222
Coordination (*see* Conjoining)
Correlatives, 135–36, 286 (*see also* Coordinating conjunctions)
Count nouns, 25–26, 286
Crossover functions (*see also* Category shift)
  of clause types, 174–76
  of words, 128–30, 139
Cross-referencing rule, 19, 286

D

Dangling modifiers, 70, 286 (*see also* Dangling participles)

Dangling participles, 63, 70, 286
Declarative clauses, 159, 160–61, 286
  as questions, 174
  as softened commands, 175
Declarative questions, 286 (*see* Echo questions)
Deferred prepositions, 119–20, 122, 166, 286
Definite article, 27–28, 286
Deletion, 245–46, 286
Demonstrative pronouns, 88–89, 286
Demonstratives, 286
  as determiners, 28, 88
  as pronouns, 88–89
Dependent clauses, 18, 199, 286 (*see also* Subordinate clauses)
Derivational affixes, 286
  of adjectives, 100
  of adverbs, 107
  of nouns, 24
  of verbs, 43
Determiner system, 27–30 (*see also* Noun phrases)
  determiners, 27–29, 80
  postdeterminers, 29–30
  predeterminers, 29–30
Determiners, 27–29, 287 (*see also* Determiner system and Noun phrases)
  articles, 27–28
  demonstratives, 27–28
  interrogatives, 164
  possessive noun phrases, 28–29
  possessive pronouns, 27
  quantities, 27
Dialects, 3, 246–49, 287 (*see also* Regional variation and African American Vernacular English)
Dictionaries:
  eighteenth century, 6
  making of, 11–12, 20–21
  purposes of, 11–12, 21, 249
Direct objects, 32–33, 287 (*see also* Noun phrases, functions)
  as component of a verb phrase, 61
  gerundive phrases as, 37–38
  grammatical vs. logical, 148–50
  identifying, 32–33
  infinitival phrases as, 37–38
  in the passive voice, 145–50
Discourse functions, 287
  crossover, 174–76
  declarative, 159, 160–61
  defined, 159–60
  exclamative, 160
  imperative, 160
  interrogative, 159, 161–72
  minor, 177
Disjunctions, 66, 135, 198, 287
Dissimilation, 254, 287

Ditransitive verbs, 70
Double negatives, 7, 8, 185–86, 191–92 (*see also* Negation)
Dual number, 93–94
Dummy subjects (*see* Subjects, placeholders)

E

Echo questions, 287
  wh-, 170
  yes-no, 170
Eighteenth century grammarians, 5, 10, 13, 14, 81–82, 84, 95, 111, 122, 185–86
Ellipsis, 137–38, 140, 222–23, 287
Embedded questions, 171, 287
Embedded sentences, 287 (*see also* Subordinate clauses)
Endingless adverbs (*see* Flat adverbs)
Epenthesis, 245, 254, 287
Exceptions (*see* Irregularities)
Exclamative clauses, 160, 172–73, 177, 287
Expletives (*see* Subjects, placeholders)
Extraposed subject noun clauses, 205–6, 287 (*see also* Noun clauses)
Eye dialect, 255

F

Finite verb forms, 46, 287
Finite verb phrases, 60–62, 63, 287
Flat adverbs, 107–8, 112, 129, 287
Free morphemes, 236, 287 (*see also* Morphemes)
Fricatives, 241–42, 287
Full passives, 287 (*see* Passive voice)
Future tense, 54–55
  formation, 55
  meaning, 55

G

Gender, 287
  agreement with gender unspecific antecedents, 81, 85
  natural vs. grammatical, 79
  of pronouns, 79, 81–82
Get- passives, 148, 153
Gerundive noun phrases, 36–38, 287
Gerunds, 36, 44, 287 (*see also* Verbal nouns)
Good vs. *well*, 108–10
Gove, Philip, 111, 21
Gradable adjectives, 17, 102, 130, 192, 287
  (*see also* Adjectives)

Grammatical direct object, 148–50
Grammatical object of a preposition, 148–50
Grammatical relations (*see also* Noun
     phrases, functions)
  in active vs. passive clauses, 145–50
  of noun phrases, 30–36
Grammatical relations ambiguity, 131, 287
     (*see also* Ambiguity)
Grammatical subject, 148–50, 287 (*see also*
     Grammatical relations)
Greek, Classical, 5, 6, 10
Grouping ambiguity, 131, 288 (*see also* Am-
     biguity)
Groupings (*see* Constituents)

H

Head of a constituent, 18, 288
  of an adjective phrase, 104
  of an adverb phrase, 111
  of a noun phrase, 27, 29, 67
  of a relative clause, 89, 207–8
  of a verb phrase, 60
Helping verbs, 44, 51–54, 288 (*see also* Aux-
     iliary verbs and Modal verbs)
  with adverbs, 190
  in negation, 183–84, 188–89
  in questions, 162–63, 165–66, 168–69
Historical present, 57
Homonyms, 132, 252 (*see also* Lexical am-
     biguity)
Human nouns, 24–25, 288
Hypercorrection, 3, 288
  with adjectives and adverbs, 110
  in pronoun usage, 83
Hypothetical *if-then* statements, 53, 54

I

Ideographic writing systems (*see* Word-
     writing systems)
Imperative clauses, 160, 172, 174, 177, 288
Inanimate nouns, 24–25, 288
Indefinite articles, 27–28, 288
Indefinite negation, 184–86, 288
Indefinite pronouns, 93, 288
  negation of, 184–86
Independent clauses, 18, 199, 288 (*see also*
     Clauses, main)
Indirect object inversion, 33–34, 35, 39, 288
Indirect objects, 33–34, 122, 288 (*see also*
     Noun phrases, functions)
  inverted, 33–34
  as subjects of passive sentences, 150
Infinitival noun phrases, 37–38, 288

Infinitives, 288
  bare, 44
  split, 39
  as verbal nouns, 36–38
Inflectional affixes, 288
  of adjectives, 100–101
  of nouns, 22–23, 244–45
  of verbs, 44–46, 245, 247–48
Intensifiers, 102, 106, 288 (*see also* Adverbs)
International Phonetic Association, 253
Interrogative clauses, 159, 161–72, 288
  minor question types, 170–72
  rhetorical questions, 174
  as softened commands, 174–75
  tag questions, 167–70
  wh- questions, 163–67
  yes-no questions, 161–63
Interrogative pronouns, 91–92, 95, 164–67,
     288
Interrogative words (*see* wh- words)
Intransitive verbs, 17, 60–61, 288 (*see also*
     Verb phrases)
Inverted indirect objects (*see* Indirect ob-
     ject inversion)
Irregularities:
  of auxiliary verbs, 51–52, 65
  of noun plurals, 22–23
  of past participles, 45, 47–49
  of past tense verbs, 46–49
  of reflexive pronouns, 87–88
  relevance to language change, 13, 19,
     49–50
  usage problems, 49–50
It:
  as a personal pronoun, 80–81
  as a subject placeholder, 31–32, 78

J

Johnson, Samuel, 6 (*see also* Dictionaries,
     making of)

L

Language academies (*see* Academies, lan-
     guage)
Language change, 12–14, 19, 38, 49–50,
     93–94, 176, 191–93
  and lexical categories, 138–30
  and spelling, 245–46, 250
Latin, Classical, 5, 6, 10
Less vs. *fewer*, 26, 39 (*see also* Count nouns
     and Noncount nouns)
Lexical ambiguity, 132–33, 288 (*see also*
     Ambiguity)

Lexical categories, 17, 288
  crossover, 128–30
Lexicography (*see* Dictionaries, making of)
Lie vs. *lay*, 50
Linguistic ambiguity (*see* Ambiguity)
Linguistic insecurity, 3, 14, 49–50
Linguistics, 7
Linking verbs, 60, 61–62, 110, 288
Literary Standard American English, 2
Logical direct object, 149–50, 288 (*see also*
    Grammatical relations)
Logical subject, 148–50, 288 (*see also* Gram-
    matical relations)
Logical vs. grammatical relations, 148–50
Lowth, Robert, 6

M

Main clauses, 289 (*see also* Clauses, main)
Main verbs, 50, 289
Manner of articulation (*see* Obstruction of
    air)
Mass nouns, 289 (*see also* Noncount nouns)
Matrix clauses, 289 (*see also* Clauses, main)
Metathesis, 246, 289
Middle voice, 154–55
Minor question types, 170–72
  echo, 170
  embedded, 171
Modal auxiliaries, 289 (*see also* Modal verbs)
Modal verbs, 51–53, 289
  meanings, 52–53
  uses, 53–54
  will vs. *shall*, 69
Modification ambiguity, 131, 289 (*see also*
    Ambiguity)
Modifiers, 18, 289
  of adjectives, 102–5
  of adverbs, 106
  dangling, 70
  of nouns, 27–30, 100–104
  of verbs, 60, 105–6
Morphemes, 236–38, 253, 289
Morphological rules, 244–46
Morphology, 236–38, 244–46, 289 (*see also*
    Word construction)
Multiple meanings (*see* Ambiguity)

N

Nasal sounds, 240–41, 289
Negation:
  of adjectives and adverbs, 187–88, 192
  of compounds, 188–89
  defined, 182

double and multiple, 185–86, 191–92
of indefinites, 184–86
of nouns, 186–87
partial, 169, 189–91
in tag questions, 168–69
of verbs, 182–84, 192
Negative clauses, 289 (*see also* Clauses, af-
    firmative vs. negative)
Negative prefixes, 187–88
Newman, Edwin, 8, 9
Noncount nouns, 25–26, 289
Nonfinite verb forms, 44–45, 51, 289
Nonfinite verb phrases, 62–63, 289 (*see also*
    Absolutes)
Nongradable adjectives, 17, 102, 130, 192,
    289 (*see also* Adjectives)
Nonhuman nouns, 24–25, 289
Nonproductive morphemes, 253
Nonrestrictive relative clauses, 209–11, 289
Nonstandard usage (*see* Usage)
Noun clauses, 202–7, 209, 289
  complement, 206–7
  extraposed subject, 205–6
  object, 202–3
  subject, 203–4
Noun negation, 186–87, 289
Noun phrases, 18, 27–38, 289
  composition, 27–30
  compound, 66, 133–34
  functions, 30–36, 202–4
    complement, 35–36
    direct object, 32–33, 37–38, 61, 148–50
    indirect object, 33–34, 35, 150
    object of a preposition, 34–35, 36, 149
    in the passive voice, 148–50
    subject, 18, 30–32, 37, 64–68, 148–50
  gerundive, 36–38
  infinitival, 36–38
  in subject-verb agreement, 164–68
Noun subcategories, 24–26
  abstract vs. concrete, 24
  animate vs. inanimate, 24–25
  collective, 68
  common vs. proper, 17, 24
  count vs. noncount, 25–26
  human vs. nonhuman, 24–25
Nouns, 17, 22–26, 289 (*see also* Noun sub-
    categories and Noun phrases)
  definition, 22
  derivational suffixes , 24
  inflectional suffixes , 22–23, 244–45, 248
  negation, 186–87
Number, 289
  of nouns, 22–23, 71, 244–45
  of *none*, 95
  of pronouns, 79, 92–94
  of verbs, 46
  in subject-verb agreement, 64–66, 71

Numbers: (*see also* Postdeterminers)
  cardinal, 30
  ordinal, 30

O

Object complements, 290
  adjectives, 103–4
  noun phrases, 35
Object noun clauses, 290 (*see also* Noun
    clauses)
Object pronouns (*see also* Case)
  forms, 79
  usage, 82–85
Objective case (*see* Case)
Objects:
  direct, *(see* Direct objects)
  indirect, *(see* Indirect objects)
  of prepositions, *(see* Objects of a preposi-
    tion)
Objects of a preposition, 34–35, 117, 290 (*see
    also* Noun phrases, functions)
  gerundive phrases as, 37
  identifying, 34
  moved from the preposition, 119–20
Obstruction of air, 240, 241–42 (*see also*
    Consonants)
Oral sounds, 240, 290
Ordinal numbers (*see* Numbers)
The Oxford English Dictionary, 7 (*see also*
    Dictionaries)

P

Paraphrases, 131, 290 (*see also* Ambiguity)
Parts of speech, 290 (*see also* Lexical cate-
    gories)
Partial negation, 290 (*see also* Negation)
Participial phrases, 278 (*see also* Nonfinite
    verb phrases)
Participles:
  past, 44–45, 153–54, 155, 290
  present, 44, 291
Particle-movement, 121, 290
Particles, 120–21, 290
Passive test, 32, 33, 39, 147
Passive voice, 146–55, 290
  full, 147–50, 152–53
  functions of, 150–51
  get- passives, 148, 153
  with imperatives, 172
  with noun clauses, 215
  with progressive tenses, 155
  in questions, 163, 167

and structural ambiguity, 153–54
  truncated, 151–54
Past participles (*see* Participles, past)
Past tense:
  in African American Vernacular English,
    249
  of *be,* 52, 65, 69
  compared to present perfect tense, 70
  inflectional suffix, 46
  irregularities, 46–50
  meaning, 55
  pronunciation of, 253–54
Perfect aspect, 57–58, 290
Perfect tenses, 56–57, 70, 290
Person, 64, 290
  defined, 64
  of pronouns, 79
  relevance to subject-verb agreement, 64
Personal pronouns, 79–85, 290
  case of, 79
  distinguishing characteristics, 79–80
  usage, 81–85, 95
Phonemes, 239–40, 290
Phonemic writing system, 251 (*see also* Al-
    phabets)
Phonetics, 240–44, 253, 290 (*see also* Pro-
    nunciation)
Phonology, 237, 238–40, 290
Phrases, 16, 18
  adjective, 104–6
  adverb, 111
  noun, 27–38
  prepositional, 34–35, 117–20
  verb, 60–64
Place of articulation, 242–43, 253, 290
Plural nouns, 22, 244–45 (*see also* Number)
  in African American Vernacular English,
    249
  irregular, 22–23
Possessive case, 291 (*see also* Possessive
    nouns and Possessive pronouns)
  nouns, 22–23
  pronouns, 79–80
Possessive noun phrases, 28–29, 291
Possessive nouns, 22–23
  in African American Vernacular English,
    248
  pronunciation, 244–45
  punctuation, 23
Possessive pronouns, 291
  as determiners, 28, 80
  long forms, 80
  short forms, 79–80
Postdeterminers, 29–30 (*see also* Deter-
    miner system)
Predeterminers, 29 (*see also* Determiner
    system)

Predicates, 18, 31, 64, 291 (*see also* Verb phrases)
Predicate adjectives, 103–4, 291
Predicate nominals, 291 (*see also* Complements, subject)
Predicate nominatives, 291 (*see* Complements, subject)
Prefixes, 291 (*see* Affixes)
Prepositional phrases, 34–35, 117–20, 291
    composition, 117
    embedded, 122
    functions, 118
Prepositions, 17, 34–35, 116–17, 291 (*see also* Prepositional phrases)
    compared to particles, 120–21
    deferred, 119–20, 122
Present participles (*see* Participles, present)
Present tense:
    African American Vernacular English, 247–48
    of *be*, 52, 65, 69
    formation, 55, 66
    inflectional suffix, 46, 245
    meaning, 54–55
    pronunciation, 245
Printing press, 4–5
Pro-forms, 78, 119, 123
Progressive aspect, 56, 291
Progressive tenses, 56–57, 59, 155
Pronouns, 17, 77–99, 291
    antecedents of, 77–78
    definition, 77
    demonstrative, 88–89
    indefinite, 93, 184–86, 192
    interrogative, 91–92, 95
    reciprocal, 88
    reflexive, 17, 85–88, 95
    relative, 17, 89–91, 95, 207–8
    universal, 92–93
Pronunciation, 237–49, 253–55 (*see also* Phonetics)
    allomorphs of the same morpheme, 237–38, 244–46
    allophones of the same phoneme, 238–40
    noun plurals, 244–45
    noun possessives, 245
    present tense verbs, 245
    and spelling, 250–52
Proper nouns, 17, 24, 291
Punctuation:
    adverbial clauses, 201
    conjoined clauses, 135–36
    conjoined constituents, 140
    interrogatives, 162
    sentence adverbs vs. coordinating conjunctions, 135–36

Q

Quantities, 291
    with count and noncount nouns, 25–26, 29
    as determiners, 28
    as indefinite pronouns, 93
    as postdeterminers, 30
Questions (*see* Interrogative clauses and Wh- words)

R

r-less dialects, 3, 247
Reciprocal pronouns, 291 (*see also* Pronouns, reciprocal)
Reduced passives (*see* Passive voice, truncated)
Reduced relative clauses, 211–14, 291
Reflexive pronouns, 291
    forms, 85–87
    nonstandard, 87–88
    use, 85–86
Reflexive verbs, 95
Regional variations, 3, 246–47, 250–51
Regularization:
    of plural nouns, 22–23
    of reflexive pronouns, 87–88
    of verbs, 49–50
Relative clauses, 89–91, 119–20, 128, 207–14, 291
    reduced, 211–15
    restrictive vs. nonrestrictive, 209–11
Relative nominals, 214
Relative pronouns, 89–91, 291
    forms, 89–90, 207
    omission of, 211–12
    usage, 90–91, 95, 207, 210
Restrictive relative clauses, 209–11, 291
Rhetorical questions, 174
Roots, 16, 188, 235–37, 292

S

Sentence adverbs, 106, 292
    compared to coordinating conjunctions, 135–36
Sentence building (*see also* Sentences and Clauses)
    through coordination, 197–99
    through subordination, 199–215
Sentence diagramming (*see* Tree diagrams)
Sentences, 16, 18, 191, 292 (*see also* Sentence building)
    distinguished from clauses, 191, 197

Sentences, *(cont.)*
    complex, 224–28, 229
    compound, 222–24, 229
    compound-complex, 228–30
    simple, 220–22, 229
Sexist language, 81
Shakespeare, William, 7, 13, 69, 176,
        193
Simon, John, 8, 9
Simple future *(see* Future tense)
Simple past *(see* Past tense)
Simple present *(see* Present tense)
Simple sentences, 220–22, 292
Simple tenses, 55–56, 292
Singular *(see also* Number)
    of nouns, 22
    of verbs, 46, 55
Sociolinguistics, 7
Softened imperatives, 53–54, 174–76
Sound production *(see* Phonetics)
Sound sequences, 237–40, 243–46, 249 *(see
        also* Pronunciation)
Speech production *(see* Phonetics)
Spelling, 244, 245–46, 249–52
    British vs. American, 254
Spelling reform, 250–52
Spelling systems, 254–55
Split infinitives, 39, 292
Standard American English, 2–3, 7–8, 14,
        292
Standard English:
    American, 2–3, 7–8, 14, 292
    other than American, 12, 254
    spelling, 249–52
Stops, 241–42, 292
Structural ambiguity, 130–33, 292 *(see also*
        Ambiguity)
Subcategories:
    adjectives, 102
    adverbs, 105–7
    crossovers of, 130
    nouns, 24–26
    pronouns, 79–93
    verbs, 50–54, 60–62
Subclasses *(see* Subcategories)
Subject complements:
    adjectives, 103–4
    confusion with adverbs, 110
    gerundive phrases, 37–38
    identifying, 35
    infinitival phrases, 37–38
    noun phrases, 35
Subject noun clauses, 292 *(see* Noun
        clauses)
Subject pronouns *(see also* Case)
    forms, 79–80
    usage, 82–85

Subjects, 18, 292 *(see also* Noun phrases,
        functions)
    in active vs. passive voice, 148–50
    compound, 66, 83
    gerundive phrases, 37
    grammatical vs. logical, 148–50
    identifying, 30–31, 82
    infinitival phrases, 37
    placeholders, 31, 32, 67, 78
    in subject-verb agreement, 64–68
    of tag questions, 168
Subjective case, 292 *(see* Case)
Subject-verb agreement, 64–68, 292
    with *be*, 65
    compound noun phrases, 66
    disjunctions, 66
    displaced subjects, 67
    passive voice, 149
    present tense, 66
    usage problems, 66–68, 70
Subordinate clauses, 199–215, 292 *(see also*
        Sentences, complex and Sentences,
        compound-complex)
    adverbial, 201–2
    identifying, 199–201
    noun, 202–7
    relative, 207–14
Subordinating conjunctions, 201–2, 292 *(see
        also* Adverbial clauses)
Subordination, 199–215, 292 *(see also* Sub-
        ordinate clauses)
Suffixes *(see* Affixes)
Superlatives:
    of adjectives, 100–102, 112, 292
    of adverbs, 107, 109
    double, 111
Superordinate clauses, 292 *(see also* Clauses,
        main)
Swift, Jonathan, 6, 247
Syllabaries, 254
Syntax, 230, 292

T

Tag questions, 167–70, 176–77, 292
Tenses of verbs, 293 *(see also* Verb
        tense)
There:
    as a placeholder, 31, 32, 67
    as a pro-form, 78
Times of verbs, 54–56, 293 *(see also* Verb
        tense)
Transitive verbs, 17, 60–61, 145, 293 *(see also*
        Direct objects)
Tree diagrams, 221–30

Truncated passives, 151–54, 293 (*see also* Passive voice)

U

Universal pronouns, 92–93, 293
Usage:
  ain't, 183, 191
  aren't I?, 169
  be, 52, 65, 69
  can vs. *may*, 53
  dangling modifiers, 70
  dangling participles, 63, 70
  deferred prepositions, 119–20, 122
  double and multiple negation, 7–8,
    185–86, 191–92
  hypothetical *if-then* statements, 69
  less vs. *fewer*, 26, 39
  lie vs. *lay*, 50
  past tense and past participles, 49–50
  personal pronouns, 81–85, 94
  reflexive pronouns, 87–88, 95
  shall vs. *will*, 53
  split infinitives, 39
  subject-verb agreement, 65–68
  well vs. *good*, 108–10
  which vs. *that*, 207, 210
  who vs. *whom*, 89–92

V

Verb negation, 182–84, 192, 293
Verb phrases, 18, 60–64, 293
  finite, 60–62
  nonfinite, 62–63
Verb tense, 46, 54–60
  complex, 56–60
    with perfect aspect, 57–58, 59
    with progressive aspect, 56–57, 59
  simple, 55–56
Verbal nouns, 36, 293
  gerunds, 36–38
  infinitives, 36–38
Verbs, 17, 43–76
  helping, 44
  intransitive, 17, 60
  linking, 60–62, 110

  main, 50
  transitive, 17, 60–61
Vocal cords, 240–41
Voice:
  active, 145–46
  defined, 145
  middle, 154–55
  passive, 146–55
Voiced sounds, 241, 293
Voiceless sounds, 241, 293
Vowels, 243, 293

W

Webster, Noah, 6, 254
Webster's Third New International Dictionary,
  11–12 (*see also* Dictionaries)
Wh- questions, 163–67, 293
Wh- words, 164, 167 (*see also* Interrogative
  pronouns)
Whiz-deletion, 212, 214, 293
Who vs. *whom*, 89–92
Word classes, 293 (*see also* Lexical categories)
Word construction, 235–38, 244–47
Word order, 19
  active vs. passive clauses, 145–48
  adverbial clauses, 201
  adverbs, 190
  declarative clauses, 160–61
  compound negation, 188–89
  embedded questions, 171
  exclamatives, 173
  relative clauses, 208
  relative pronouns, 207
  tag questions, 168–69
  wh- questions, 164–66
  yes-no questions, 162
Words, 293 (*see also* Lexical categories)
  construction of, 235–38
  as a level of structure, 16–17
Word-writing systems, 255
Writing systems (*see* Spelling)

Y

Yes-no questions, 161–63, 191, 293